AF679962

Water Conservation and Management: Principles and Practices

Water Conservation and Management: Principles and Practices

Vincent Ford

STATES
ACADEMIC PRESS
www.statesacademicpress.com

Published by States Academic Press,
109 South 5th Street,
Brooklyn, NY 11249, USA
www.statesacademicpress.com

Water Conservation and Management: Principles and Practices
Vincent Ford

© 2022 States Academic Press

International Standard Book Number: 978-1-63989-563-2 (Hardback)

This book contains information obtained from authentic and highly regarded sources. All chapters are published with permission under the Creative Commons Attribution Share Alike License or equivalent. A wide variety of references are listed. Permissions and sources are indicated; for detailed attributions, please refer to the permissions page. Reasonable efforts have been made to publish reliable data and information, but the authors, editors and publisher cannot assume any responsibility for the validity of all materials or the consequences of their use.

Trademark Notice: Registered trademark of products or corporate names are used only for explanation and identification without intent to infringe.

Cataloging-in-Publication Data

Water conservation and management : principles and practices / Vincent Ford.
p. cm.
Includes bibliographical references and index.
ISBN 978-1-63989-563-2
1. Water conservation. 2. Water quality management. 3. Water-supply--Management.
4. Water--Pollution. I. Ford, Vincent.

TD388 .W38 2022

333.911 6--dc23

Table of Contents

Preface

It is with great pleasure that I present this book. It has been carefully written after numerous discussions with my peers and other practitioners of the field. I would like to take this opportunity to thank my family and friends who have been extremely supporting at every step in my life.

The practice of conserving water using strategies, policies and activities to sustainably manage the fresh water resources is referred to as water conservation. It focuses on protecting the hydrosphere, and ensuring current and future demand of water. Rainwater harvesting is one of the most prominent strategic measure to conserve water. Water management is the management of water resources through planning, developing and distributing water resources under established policies and regulations. Water treatment, water resource management, irrigation management, and water table management are the focus areas of water management. Key strategies involved in water protection are avoiding damage of water quality, improvement of water management practices and checking for water loss. This textbook elucidates the concepts and innovative models around prospective developments with respect to water conservation and management. The topics covered herein deal with the core aspects of water conservation and management. This book will serve as a reference to a broad spectrum of readers.

The chapters below are organized to facilitate a comprehensive understanding of the subject:

Chapter - Water Resources

The natural resources of water that are useful to society for agricultural, industrial, household, recreational and environmental activities are referred to as water resources. The sources of water include lakes, rivers, wetlands, oceans, groundwater, wells, etc. This is an introductory chapter which will briefly introduce all the aspects related to water resources and its types.

Chapter - Water Scarcity

The lack of freshwater resources to fulfil the demands of water usage in a region is known as water scarcity. The major causes of water scarcity are water pollution and overpopulation. This chapter has been carefully written to provide an easy understanding of the subject of water scarcity.

Chapter - Water Conservation

Water conservation refers to the policies, strategies and activities performed for the sustainable management and protection of the water resources. Some of the methods used for water conservation are rainwater harvesting, desalination, drip irrigation, etc. The topics elaborated in this chapter will help in gaining a better perspective about the different methods and techniques used for water conservation.

Chapter – Water Resource Management

The activity of planning, developing, distributing and managing water resources to ensure their optimum use is referred to as water management. It is primarily concerned with water resource management, irrigation management, integrated urban water management and storm water management. All these diverse aspects of water management have been carefully analyzed in this chapter.

Chapter – Water Treatment

The process used to improve the quality of water to make it potable for various uses is referred to as water treatment. Water treatment is done through the processes of primary, secondary and advanced treatment. This chapter closely examines these processes used in water treatment for an extensive understanding of the subject.

Vincent Ford

1

Water Resources

The natural resources of water that are useful to society for agricultural, industrial, household, recreational and environmental activities are referred to as water resources. The sources of water include lakes, rivers, wetlands, oceans, groundwater, wells, etc. This is an introductory chapter which will briefly introduce all the aspects related to water resources and its types.

WATER

Water is a substance composed of the chemical elements hydrogen and oxygen and existing in gaseous, liquid, and solid states. It is one of the most plentiful and essential of compounds. A tasteless and odourless liquid at room temperature, it has the important ability to dissolve many other substances. Indeed, the versatility of water as a solvent is essential to living organisms. Life is believed to have originated in the aqueous solutions of the world's oceans, and living organisms depend on aqueous solutions, such as blood and digestive juices, for biological processes. In small quantities water appears colourless, but water actually has an intrinsic blue colour caused by slight absorption of light at red wavelengths.

Although the molecules of water are simple in structure (H_2O), the physical and chemical properties of the compound are extraordinarily complicated, and they are not typical of most substances found on Earth. For example, although the sight of ice cubes floating in a glass of ice water is commonplace, such behaviour is unusual for chemical entities. For almost every other compound, the solid state is denser than the liquid state; thus, the solid would sink to the bottom of the liquid. The fact that ice floats on water is exceedingly important in the natural world, because the ice that forms on ponds and lakes in cold areas of the world acts as an insulating barrier that protects the aquatic life below. If ice were denser than liquid water, ice forming on a pond would sink, thereby exposing more water to the cold temperature. Thus, the pond would eventually freeze throughout, killing all the life-forms present.

Water occurs as a liquid on the surface of Earth under normal conditions, which makes it invaluable for transportation, for recreation, and as a habitat for a myriad of plants and animals. The fact that water is readily changed to a vapour (gas) allows it to be transported through the atmosphere from the oceans to inland areas where it condenses and, as rain, nourishes plant and animal life.

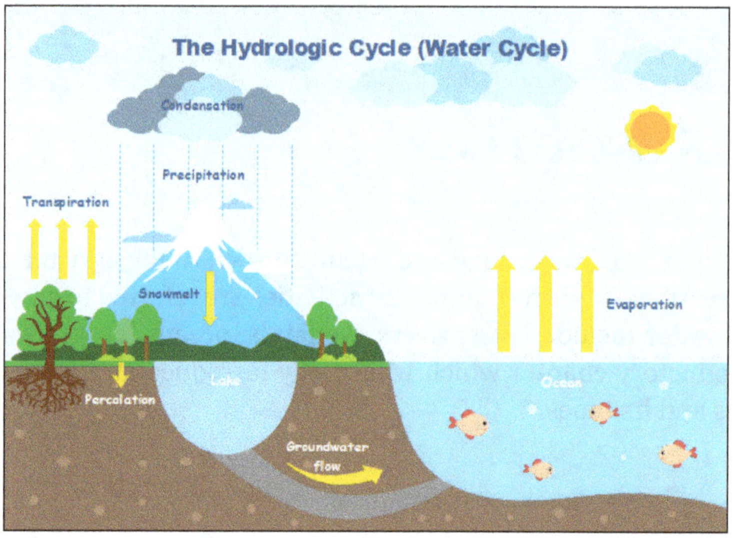

In the hydrologic cycle, water is transferred between the land surface, the ocean, and the atmosphere. The numbers on the arrows indicate relative water fluxe.

Because of its prominence, water has long played an important religious and philosophical role in human history. In the 6th century bce, Thales of Miletus, sometimes credited for initiating Greek philosophy, regarded water as the sole fundamental building block of matter:

> It is water that, in taking different forms, constitutes the earth, atmosphere, sky, mountains, gods and men, beasts and birds, grass and trees, and animals down to worms, flies and ants. All these are different forms of water. Meditate on water.

Two hundred years later, Aristotle considered water to be one of four fundamental elements, in addition to earth, air, and fire. The belief that water was a fundamental substance persisted for more than 2,000 years until experiments in the second half of the 18th century showed that water is a compound made up of the elements hydrogen and oxygen.

The water on the surface of Earth is found mainly in its oceans (97.25 percent) and polar ice caps and glaciers (2.05 percent), with the balance in freshwater lakes, rivers, and groundwater. As Earth's population grows and the demand for fresh water increases, water purification and recycling become increasingly important. Interestingly, the purity requirements of water for industrial use often exceed those for human consumption. For example, the water used in high-pressure boilers must be at least 99.999998 percent pure. Because seawater contains large quantities of dissolved salts, it must be desalinated for most uses, including human consumption.

The Hoover Dam on the Colorado River at the border of Nevada and Arizona demonstrates how natural resources of water can be harnessed for a variety of purposes, including human consumption, irrigation, and industry.

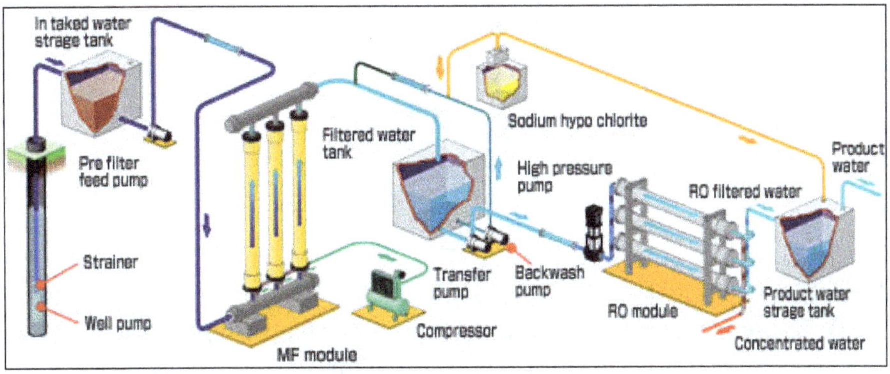

Water treatment systems are important for desalinating seawater so it can be used for human consumption and for purifying water for industrial use.

Structure of Water

Liquid Water

The water molecule is composed of two hydrogen atoms, each linked by a single chemical bond to an oxygen atom. Most hydrogen atoms have a nucleus consisting solely of a proton. Two isotopic forms, deuterium and tritium, in which the atomic nuclei also contain one and two neutrons, respectively, are found to a small degree in water. Deuterium oxide (D_2O), called heavy water, is important in chemical research and is also used as a neutron moderator in some nuclear reactors.

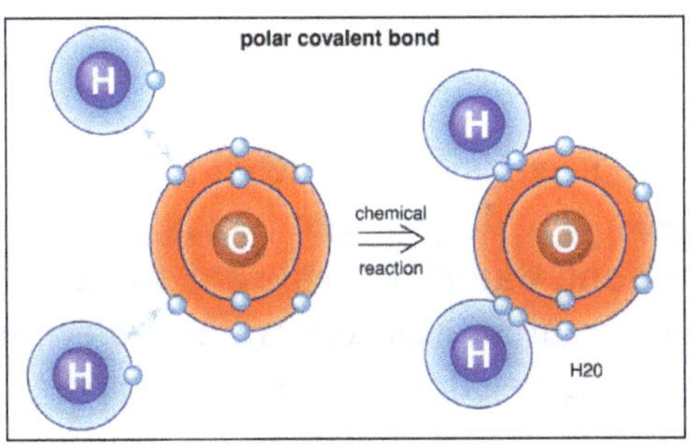

A water molecule is made up of two hydrogen atoms and one oxygen atom. A single oxygen atom contains six electrons in its outer shell, which can hold a total of eight electrons. When two hydrogen atoms are bound to an oxygen atom, the outer electron shell of oxygen is filled.

Although its formula (H_2O) seems simple, water exhibits very complex chemical and physical properties. For example, its melting point, 0 °C (32 °F), and boiling point, 100 °C (212 °F), are much higher than would be expected by comparison with analogous compounds, such as hydrogen sulfide and ammonia. In its solid form, ice, water is less dense than when it is liquid, another unusual property. The root of these anomalies lies in the electronic structure of the water molecule.

The water molecule is not linear but bent in a special way. The two hydrogen atoms are bound to the oxygen atom at an angle of 104.5°.

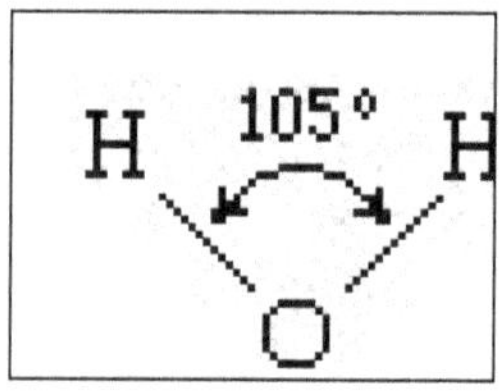

The O–H distance (bond length) is 95.7 picometres (9.57 × 10–11 metres, or 3.77 × 10–9 inches). Because an oxygen atom has a greater electronegativity than a hydrogen atom, the O–H bonds in the water molecule are polar, with the oxygen bearing a partial negative charge (δ–) and the hydrogens having a partial positive charge (δ+).

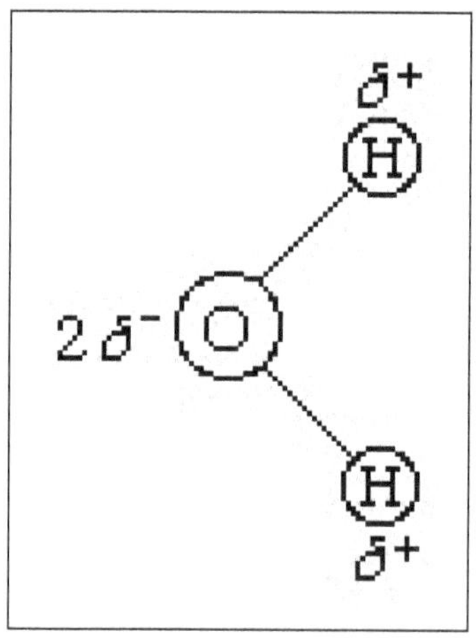

Hydrogen atoms in water molecules are attracted to regions of high electron density and can form weak linkages, called hydrogen bonds, with those regions. This means that the hydrogen atoms in one water molecule are attracted to the nonbonding electron pairs of the oxygen atom on an adjacent water molecule. The structure of liquid water is believed to consist of aggregates of water molecules that form and re-form continually. This short-range order, as it is called, accounts for other unusual properties of water, such as its high viscosity and surface tension.

Water is a polar molecule and is attracted to other polar molecules. Thus, droplets, or beads, of water form on a nonpolar surface because water molecules adhere together instead of adhering to the surface.

An oxygen atom has six electrons in its outer (valence) shell, which can hold a total of eight electrons. When an oxygen atom forms a single chemical bond, it shares one of its own electrons with the nucleus of another atom and receives in return a share of an electron from that atom. When bonded to two hydrogen atoms, the outer electron shell of the oxygen atom is filled.

The electron arrangement in the water molecule can be represented as follows.

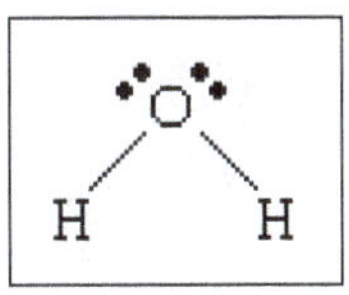

Each pair of dots represents a pair of unshared electrons (i.e., the electrons reside on only the oxygen atom). This situation can also be depicted by placing the water molecule in a cube.

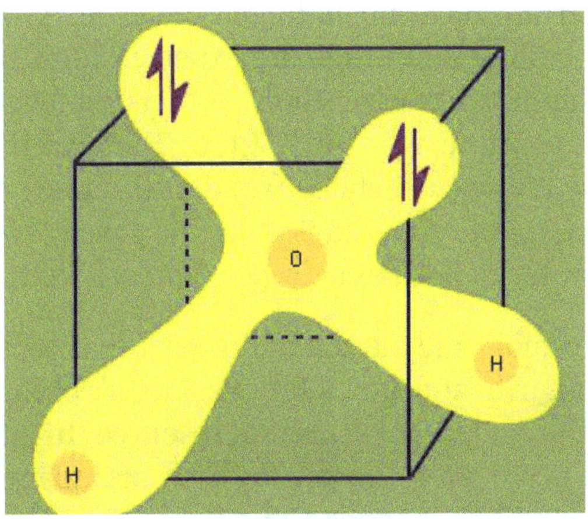

Each ↑↓ symbol represents a pair of unshared electrons. This electronic structure leads to hydrogen bonding.

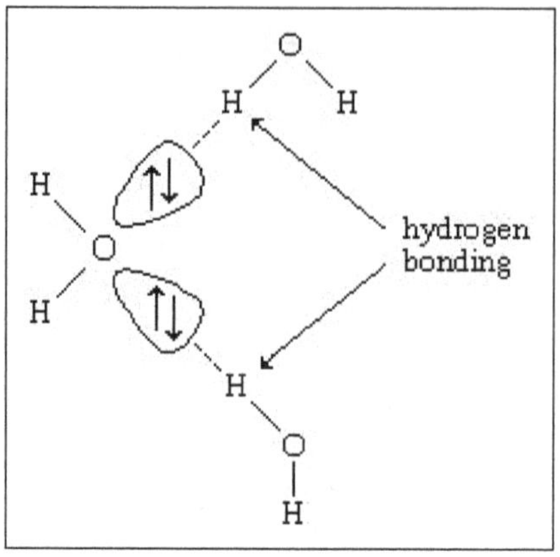

Structures of Ice

In the solid state (ice), intermolecular interactions lead to a highly ordered but loose structure in which each oxygen atom is surrounded by four hydrogen atoms; two of these hydrogen atoms are covalently bonded to the oxygen atom, and the two others (at longer distances) are hydrogen bonded to the oxygen atom's unshared electron pairs.

This open structure of ice causes its density to be less than that of the liquid state, in which the ordered structure is partially broken down and the water molecules are (on average) closer together. When water freezes, a variety of structures are possible depending on the conditions. Nine different forms of ice are known and can be interchanged by varying external pressure and temperature.

Significance of the Structure of Liquid Water

The liquid state of water has a very complex structure, which undoubtedly involves considerable association of the molecules. The extensive hydrogen bonding among the molecules in liquid water produces much larger values for properties such as viscosity, surface tension, and boiling point than are expected for a typical liquid containing small molecules. For example, based on the size of its molecules, water would be expected to have a boiling point nearly 200 °C (360 °F) lower than its observed boiling point. In contrast to the condensed states (solid and liquid) of water, which exhibit extensive association among the water molecules, its gaseous (vapour) phase contains relatively independent water molecules at large distances from each other.

The polarity of the water molecule plays a major part in the dissolution of ionic compounds during the formation of aqueous solutions. Earth's oceans contain vast amounts of dissolved salts, which provide a great natural resource. In addition, the hundreds of chemical reactions that occur every instant to keep organisms alive all take place in aqueous fluids. Also, the ability of foods to be flavoured as they are cooked is made possible by the solubility in water of such substances as sugar and salt. Although the solubility of substances in water is an extremely complex process, the interaction between the polar water molecules and the solute (i.e., the substance being dissolved) plays a major role. When an ionic solid dissolves in water, the positive ends of the water molecules are attracted to the anions, while their negative ends are attracted to the cations. This process is called hydration. The hydration of its ions tends to cause a salt to break apart (dissolve) in the water. In the dissolving process the strong forces present between the positive and negative ions of the solid are replaced by strong water-ion interactions.

When ionic substances dissolve in water, they break apart into individual cations and anions. For instance, when sodium chloride (NaCl) dissolves in water, the resulting solution contains separated Na^+ and Cl^- ions.

$$NaCl(s) \xrightarrow{H_2O} Na^+(aq) + Cl^-(aq)$$

In this equation the (s) represents the solid state, and the (aq), which is an abbreviation for aqueous, shows that the ions are hydrated—that is, they have a certain number of water molecules attached to them. As sodium chloride dissolves, four water molecules closely associate with the sodium ion. (The hydration number of Na^+ is four.) Just outside this inner hydration sphere is a region where water molecules are partially ordered by the presence of the $[Na(H_2O)_4]^+$ hydrated ion. This partially ordered region blends into "regular" (bulk) liquid water.

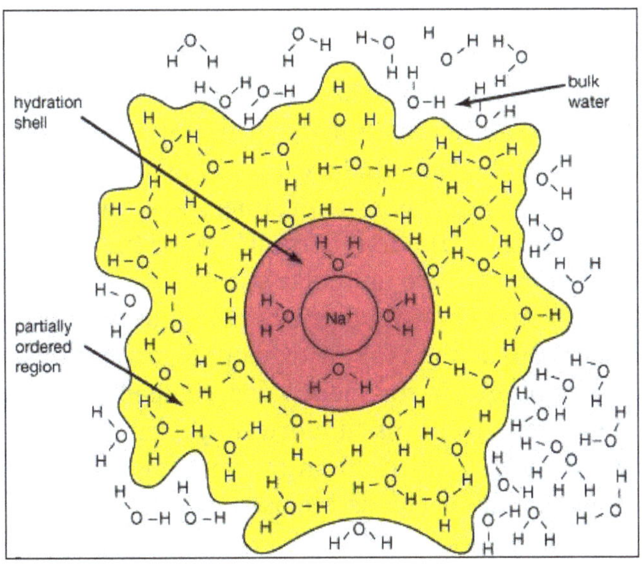

Generally speaking, the greater the charge density (the ratio of charge to surface area) of an ion, the larger the hydration number will be. As a rule, negative ions have smaller hydration numbers than positive ions because of the greater crowding that occurs when the hydrogen atoms of the water molecules are oriented toward the anion.

Many nonionic compounds are also soluble in water. For example, ethanol (C_2H_5OH), the alcoholic component of wine, beer, and distilled spirits, is highly soluble in water.

These beverages contain varying percentages of ethanol in aqueous solution with other substances. Ethanol is so soluble in water because of the structure of the alcohol molecule. The molecule contains a polar O–H bond like those in water, which allows it to interact effectively with water.

There are many substances for which water is not an acceptable solvent. Animal fat, for example, is insoluble in pure water because the nonpolar nature of fat molecules renders them incompatible with polar water molecules. In general, polar and ionic substances are soluble in water. A useful rule of thumb for determining whether two substances are likely to be miscible (i.e., will mix to form a solution) is "like dissolves like." That is, two polar substances are likely to mix to form a solution, as are two nonpolar substances.

Behaviour and Properties

Water at High Temperatures and Pressures

The characteristic ability of water to behave as a polar solvent (dissolving medium) changes when water is subjected to high temperatures and pressures. As water becomes hotter, the molecules seem much more likely to interact with nonpolar molecules. For example, at 300 °C (572 °F) and high pressure, water has dissolving properties very similar to acetone (CH_3COCH_3), a common organic solvent.

Water exhibits particularly unusual behaviour beyond its critical temperature and pressure (374 °C [705.2 °F], 218 atmospheres). Above its critical temperature, the distinction between the liquid and gaseous states of water disappears—it becomes a supercritical fluid, the density of which can be varied from liquidlike to gaslike by varying its temperature and pressure. If the density of supercritical water is high enough, ionic solutes are readily soluble, as is true for "normal" water; but, surprisingly, this supercritical fluid can also readily dissolve nonpolar substances—something ordinary water cannot do. Because of its ability to dissolve nonpolar substances, supercritical water can be used as a combustion medium for destroying toxic wastes. For example, organic wastes can be mixed with oxygen in sufficiently dense supercritical water and combusted in the fluid; the flame actually burns "underwater." Oxidation in supercritical water can be used to destroy a wide variety of hazardous organic substances with the advantage that a supercritical-water reactor is a closed system, so there are no emissions released into the atmosphere.

Physical Properties

Water has several important physical properties. Although these properties are familiar because of the omnipresence of water, most of the physical properties of water are quite atypical. Given the low molar mass of its constituent molecules, water has unusually large values of viscosity, surface tension, heat of vaporization, and entropy of vaporization, all of which can be ascribed to the extensive hydrogen bonding interactions present in liquid water. The open structure of ice that allows for maximum hydrogen bonding explains why solid water is less dense than liquid water—a highly unusual situation among common substances.

Selected physical properties of water	
Molar mass	18.0151 grams per mole
Melting point	0.00 °C
Boiling point	100.00 °C
Maximum density (at 3.98 °C)	1.0000 grams per cubic centimetre
Density (25 °C)	0.99701 grams per cubic centimetre
Vapour pressure (25 °C)	23.75 torr
Heat of fusion (0 °C)	6.010 kilojoules per mole
Heat of vaporization (100 °C)	40.65 kilojoules per mole
Heat of formation (25 °C)	-285.85 kilojoules per mole
Entropy of vaporization (25 °C)	118.8 joules per °C mole
Viscosity	0.8903 centipoise
Surface tension (25 °C)	71.97 dynes per centimetre

Chemical Properties

Acid-base Reactions

Water undergoes various types of chemical reactions. One of the most important chemical properties of water is its ability to behave as both an acid (a proton donor) and a base (a proton acceptor), the characteristic property of amphoteric substances. This behaviour is most clearly seen in the autoionization of water:

$$H_2O(l) + H_2O(l) \rightleftharpoons H_3O^+(aq) + OH^-(aq),$$

where the (l) represents the liquid state, the (aq) indicates that the species are dissolved in water, and the double arrows indicate that the reaction can occur in either direction and an equilibrium condition exists. At 25 °C (77 °F) the concentration of hydrated H^+ (i.e., H_3O^+, known as the hydronium ion) in water is 1.0×10^{-7} M, where M represents moles per litre. Since one OH^- ion is produced for each H_3O^+ ion, the concentration of OH^- at 25 °C is also 1.0×10^{-7} M. In water at 25 °C the H_3O^+ concentration and the OH^- concentration must always be 1.0×10^{-14}:

$$[H^+][OH^-] = 1.0 \times 10^{-14},$$

where $[H^+]$ represents the concentration of hydrated H^+ ions in moles per litre and $[OH^-]$ represents the concentration of OH^- ions in moles per litre.

When an acid (a substance that can produce H^+ ions) is dissolved in water, both the acid and the water contribute H^+ ions to the solution. This leads to a situation in which the H^+ concentration is greater than 1.0×10^{-7} M. Since it must always be true that $[H^+][OH^-] = 1.0 \times 10^{-14}$ at 25 °C, the $[OH^-]$ must be lowered to some value below 1.0×10^{-7}. The mechanism for reducing the concentration of OH^- involves the reaction

$$H^+ + OH^- \rightarrow H_2O,$$

which occurs to the extent needed to restore the product of $[H^+]$ and $[OH^-]$ to 1.0×10^{-14} M. Thus, when an acid is added to water, the resulting solution contains more H^+ than OH^-; that is, $[H^+] > [OH^-]$. Such a solution (in which $[H^+] > [OH^-]$) is said to be acidic.

The most common method for specifying the acidity of a solution is its pH, which is defined in terms of the hydrogen ion concentration:

$$pH = -\log [H^+],$$

where the symbol log stands for a base-10 logarithm. In pure water, in which $[H^+] = 1.0 \times 10^{-7}$ M, the pH = 7.0. For an acidic solution, the pH is less than 7. When a base (a substance that behaves as a proton acceptor) is dissolved in water, the H^+ concentration is decreased so that $[OH^-] > [H^+]$. A basic solution is characterized by having a pH > 7. In summary, in aqueous solutions at 25 °C:

neutral solution	$[H^+] = [OH^-]$	pH = 7
acidic solution	$[H^+] > [OH^-]$	pH < 7
basic solution	$[OH^-] > [H^+]$	pH > 7

When an active metal such as sodium is placed in contact with liquid water, a violent exothermic (heat-producing) reaction occurs that releases flaming hydrogen gas.

$$2Na(s) + 2H_2O(l) \rightarrow 2Na^+(aq) + 2OH^-(aq) + H_2(g)$$

This is an example of an oxidation-reduction reaction, which is a reaction in which electrons are transferred from one atom to another. In this case, electrons are transferred from sodium atoms (forming Na^+ ions) to water molecules to produce hydrogen gas and OH^- ions. The other alkali metals give similar reactions with water. Less-active metals react slowly with water. For example, iron reacts at a negligible rate with liquid water but reacts much more rapidly with superheated steam to form iron oxide and hydrogen gas.

$$3Fe(s) + 4H_2O(g) \xrightarrow{600°C} Fe_3O_4(s) + 4H_2(g)$$

Noble metals, such as gold and silver, do not react with water at all.

WATER RESOURCES

Water resources are natural resources of water that are potentially useful. Uses of water include agricultural, industrial, household, recreational and environmental activities. All living things require water to grow and reproduce.

97% of the water on the Earth is salt water and only three percent is fresh water; slightly over two thirds of this is frozen in glaciers and polar ice caps. The remaining unfrozen freshwater is found mainly as groundwater, with only a small fraction present above ground or in the air.

Fresh water is a renewable resource, yet the world's supply of groundwater is steadily decreasing, with depletion occurring most prominently in Asia, South America and North America, although it is still unclear how much natural renewal balances this usage, and whether ecosystems are threatened. The framework for allocating water resources to water users (where such a framework exists) is known as water rights.

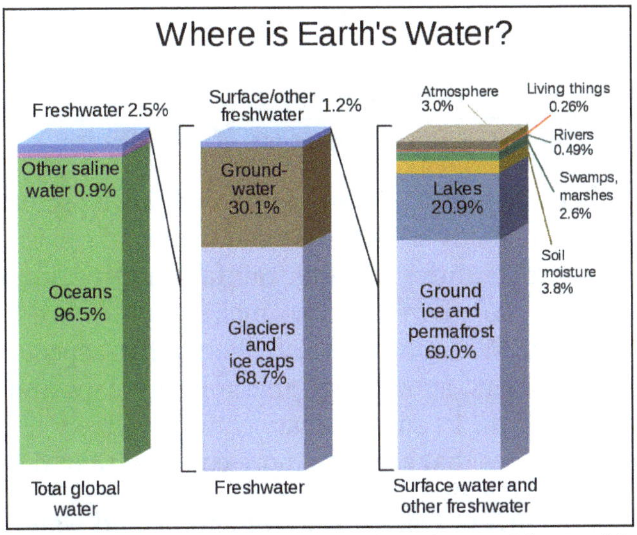

A graphical distribution of the locations of water on Earth. Only 3% of the Earth's water is fresh water. Most of it is in icecaps and glaciers (69%) and groundwater (30%), while all lakes, rivers and swamps combined only account for a small fraction (0.3%) of the Earth's total freshwater reserves.

Water resources are under major stress around the world. Rivers, lakes, and underground aquifers supply fresh water for irrigation, drinking, and sanitation, while the oceans provide habitat for a large share of the planet's food supply. Today, however, expansion of agriculture, damming, diversion, over-use, and pollution threaten these irreplaceable resources in many parts of the globe.

Providing safe drinking water for the more than 1 billion people who currently lack it is one of the greatest public health challenges facing national governments today. In many developing countries, safe water, free of pathogens and other contaminants, is unavailable to much of the population, and water contamination remains a concern even for developed countries with good water supplies and advanced treatment systems. And over-development, especially in coastal regions and areas with strained water supplies, is leading many regions to seek water from more and more distant sources.

Eastern U.S. aquifers contaminated with salt water.

Scientists widely predict that global climate change will have profound impacts on the hydrologic cycle, and that in many cases these effects will make existing water challenges worse. Rising global temperatures will alter rainfall patterns, making them stronger in some regions and weaker in others, and may make storms more frequent and severe in some areas of the world. Warming will also affect other aspects of the water cycle by reducing the size of glaciers, snowpacks, and polar ice caps and changing rates of evaporation and transpiration. In sum, climate change is likely to make many of the water-management challenges that are even more complex than they are today.

At the same time, many current trends in water supply and water quality in Europe and North America are positive. Thirty years ago, many water bodies in developed countries were highly polluted. For example, on June 22, 1969, the Cuyahoga River in Cleveland, Ohio, caught fire when sparks ignited an oily slick of industrial chemicals on its surface. Today, the United States and western European countries have reduced pollution discharges into rivers and lakes, often producing quick improvements in water quality. These gains show that when societies make water quality a priority, many polluted sources can be made usable once again. Furthermore, in the United States water consumption rates have consistently declined over the last several decades.

The Global Water Cycle

Water covers about three-quarters of Earth's surface and is a necessary element for life. During their constant cycling between land, the oceans, and the atmosphere, water molecules pass repeatedly through solid, liquid, and gaseous phases (ice, liquid water, and water vapor), but the total supply remains fairly constant. A water molecule can travel to many parts of the globe as it cycles.

Water vapor redistributes energy from the sun around the globe through atmospheric circulation. This happens because water absorbs a lot of energy when it changes its state from liquid to gas. Even though the temperature of the water vapor may not increase when it evaporates from liquid water, this vapor now contains more energy, which is referred to as latent heat. Atmospheric circulation moves this latent heat around Earth, and when water vapor condenses and produces rain, the latent heat is released.

Very little water is consumed in the sense of actually taking it out of the water cycle permanently, and unlike energy resources such as oil, water is not lost as a consequence of being used. However, human intervention often increases the flux of water out of one store of water into another, so it can deplete the stores of water that are most usable. For example, pumping groundwater for irrigation depletes aquifers by transferring the water to evaporation or river flow. Our activities also pollute water so that it is no longer suitable for human use and is harmful to ecosystems.

There are three basic steps in the global water cycle: water precipitates from the atmosphere, travels on the surface and through groundwater to the oceans, and evaporates or transpires back to the atmosphere from land or evaporates from the oceans. Figure illustrates yearly flow volumes in thousands of cubic kilometers.

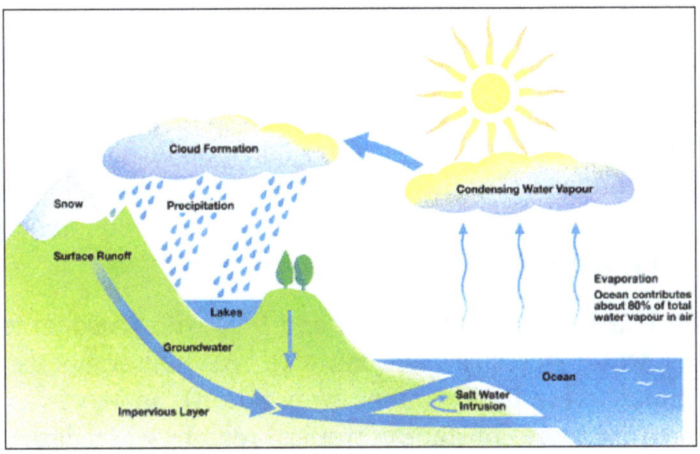

The water cycle.

Supplies of freshwater (water without a significant salt content) exist because precipitation is greater than evaporation on land. Most of the precipitation that is not transpired by plants or evaporated, infiltrates through soils and becomes groundwater, which flows through rocks and sediments and discharges into rivers. Rivers are primarily supplied by groundwater, and in turn provide most of the freshwater discharge to the sea. Over the oceans evaporation is greater than precipitation, so the net effect is a transfer of water back to the atmosphere. In this way freshwater resources are continually renewed by counterbalancing differences between evaporation and precipitation on land and at sea, and the transport of water vapor in the atmosphere from the sea to the land.

Nearly 97 percent of the world's water supply by volume is held in the oceans. The other large reserves are groundwater (4 percent) and icecaps and glaciers (2 percent), with all other water bodies together accounting for a fraction of 1 percent. Residence times vary from several thousand years in the oceans to a few days in the atmosphere.

	Surface area (million km)	Volume (million km)	Volume (%)	Equivalent depth (m)	Residence time
Oceans and seas	361	1,370	94	2,500	~4,000 years
Lakes and reservoirs	1.55	0.13	<0.01	0.25	~10 years
Swamps	<0.1	<0.01	<0.01	0.007	1-10 years
River channels	<0.1	<0.01	<0.01	0.003	~2 weeks
Soil moisture	130	0.07	<0.01	0.13	2 weeks to 50 years
Groundwater	130	60	4	120	2 weeks to 100,000 years
Icecaps and glaciers	17.8	30	2	60	10 to 1,000 years
Atmospheric water	504	0.01	<0.01	0.025	~10 days
Biospheric water	<0.1	<0.01	<0.01	0.001	~1 week

Solar radiation drives evaporation by heating water so that it changes to water vapor at a faster rate. This process consumes an enormous amount of energy—nearly one-third of the incoming solar energy that reaches Earth's surface. On land, most evaporation occurs as transpiration through plants: water is taken up through roots and evaporates through stomata in the leaves as the plant takes in CO_2. A single large oak tree can tran-spire up to 40,000 gallons per year. Much of the water moving through the hydrologic cycle thus is involved with plant growth.

Since evaporation is driven by heat, it rises and falls with seasonal temperatures. In temperate regions, water stores rise and fall with seasonal evaporation rates, so that net atmospheric input (precipitation minus evaporation) can vary from positive to negative. Temperatures are more constant in tropical regions where large seasonal differences in precipitation, such as monsoon cycles, are the main cause of variations in the availability of water. In an effort to reduce these seasonal swings, many countries have built reservoirs to capture water during periods of high flow or flooding and release water during periods of low flow or drought. These projects have increased agricultural production and mitigated floods and droughts in some regions, but as we will see, they have also had major unintended impacts on water supplies and water quality.

The hydrologic cycle is also coupled with material cycles because rainfall erodes and weathers rock. Weathering breaks down rocks into gravel, sand, and sediments, and is an important source of key nutrients such as calcium and sulfur. Estimates from river outflows indicate that some 17 billion tons of material are transported into the oceans each year, of which about 80 percent is particulate and 20 percent is dissolved. On average, Earth's surface weathers at a rate of about 0.5 millimeter per year. Actual rates may be much higher at specific locations and may have been accelerated by human activities, such as emissions from fossil fuel combustion that make rain and snowfall more acidic.

Distribution of Freshwater Resources

Freshwater accounts for only some 6 percent of the world's water supply, but is essential for human uses such as drinking, agriculture, manufacturing, and sanitation. As discussed above, two-thirds of global freshwater is found underground.

If you dig deeply enough anywhere on Earth, you will hit water. Some people picture groundwater as an underground river or lake, but in reality it is rarely a distinct water body (large caves in limestone aquifers are one exception). Rather, groundwater typically fills very small spaces (pores) within rocks and between sediment grains.

The water table is the top of the saturated zone. It may lie hundreds of meters deep in deserts or near the surface in moist ecosystems. Water tables typically shift from season to season as precipitation and transpiration levels change, moving up during rainy periods or periods of little transpiration and sinking during dry phases when the rate of recharge (precipitation minus evaporation and transpiration that infiltrates from the surface) drops. In temperate regions the water table tends to follow surface topography, rising under hills where there is little discharge to streams and falling under valleys where the water table intersects the surface in the form of streams, lakes, and springs.

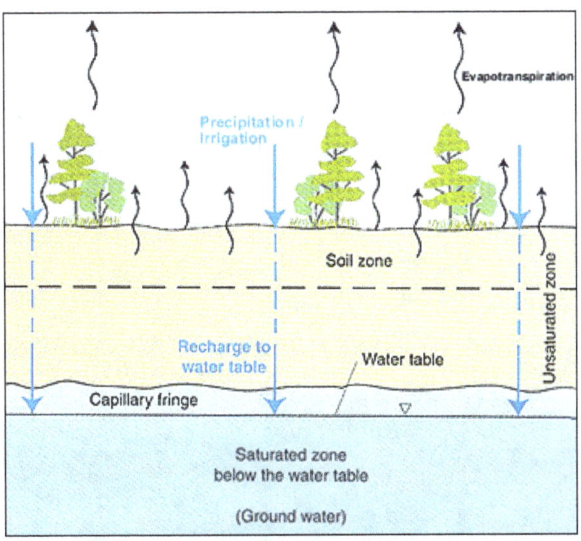

Water in the ground.

Above the water table lies the unsaturated zone, also referred to as the vadose zone, where the pores (spaces between grains) are not completely filled with water. Water in the vadose zone is referred to as soil moisture. Although air in the vadose zone is at atmospheric pressures, the soil moisture is under tension, with suctions of a magnitude much greater than atmospheric pressure.

This fluid tension is created by strong adhesive forces between the water and the solid grains, and by surface tension at the small interfaces between water and air. The same forces can be seen at work when you insert a thin straw (a capillary) into water: water rises up in the straw, forming a meniscus at the top. When the straw is thinner, water rises higher because the ratio of the surface area of the straw to the volume of the straw is greater, increasing the adhesive force lifting the water relative to the gravitational force pulling it down. This explains why fine-grained soils, such as clay, can hold water under very large suctions.

Water flows upward under suction through small pores from the water table toward plant roots when evapotranspiration is greater than precipitation. After a rainstorm, water may recharge the groundwater by saturating large pores and cracks in the soil and flowing very quickly downward to the water table.

Millions of people worldwide depend on groundwater stocks, which they draw from aquifers— permeable geologic formations through which water flows easily. Very transmissive geologic formations are desirable because water levels in wells decline little even when pumping rates are high, so the wells do not need to be drilled as deeply as in less transmissive formations and the energy costs of lifting water to the surface are not excessive. Under natural conditions many aquifers are artesian: the water they hold is under pressure, so water will flow to the surface from a well without pumping.

Aquifers may be either capped by an impermeable layer (confined) or open to receive water from the surface (unconfined). Confined aquifers are often artesian because the confining layer prevents upward flow of groundwater, but unconfined aquifers are also artesian in the vicinity of discharge areas. This is why groundwater discharges into rivers and streams. Confined aquifers are less likely to be contaminated because the impermeable layers above them prevent surface contaminants from reaching their water, so they provide good-quality water supplies.

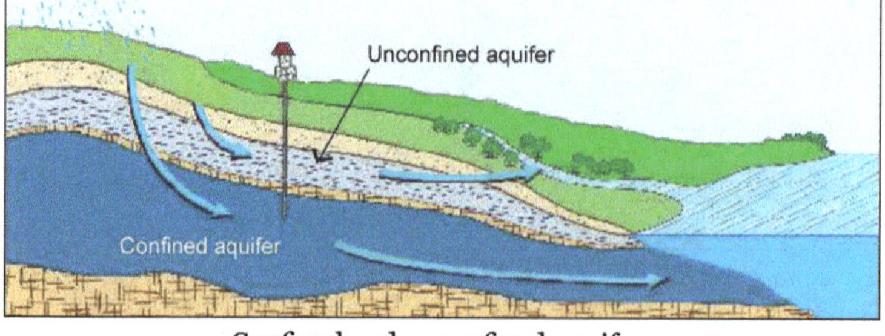

Confined and unconfined aquifer.

Water has an average residence time of thousands to tens of thousands of years in many aquifers, but the actual age of a water sample collected from a particular well will vary tremendously within an aquifer. Shallow groundwater can discharge into streams and rivers in weeks or months, but some deep groundwater is millions of years old—as old as the rocks that hold the water in their pores. Because of this distribution of residence times in aquifers, contaminants that have been introduced at the surface over the last century are only now beginning to reach well depths and contaminate drinking water in many aquifers. Indeed, much of the solute load (salt and other contaminants) that has entered aquifers due to increased agriculture and other land use changes over the last several centuries has yet to reach discharge areas where it will contaminate streams and lakes.

Ice sheets and glaciers are not always thought of as freshwater sources, but they account for a significant fraction of world reserves. Nearly 90 percent of the water in icecaps and glaciers is in Antarctica, with another 10 percent in the Greenland ice sheet and the remainder in tropical and temperate glaciers. Earth's ice sheets constantly expand and contract as the planet's climate fluctuates. During warm periods ice sheets melt and sea levels rise, with the reverse occurring when temperatures fall. Water may remain locked in deep layers of polar ice sheets for hundreds of thousands of years.

Rivers contain a relatively small share of fresh water, but the flux of water down rivers is a large part of the global hydrologic cycle and they are centrally important in shaping landscapes. Their flow erodes solid sediment and carries it toward the sea, along with dissolved minerals. These processes shape land into valleys and ridges and deposit thick layers of sediment in flood plains. Over geologic time the erosion caused by rivers balances the uplift driven by plate tectonics. Much of Earth's freshwater flow passes through several of the planet's largest rivers: the Amazon carries 15 percent of total river flow on Earth, the Congo carries 3.5 percent, and rivers that flow into the Arctic Ocean carry 8 percent. The average residence time of water in rivers is less than a year.

Groundwater Hydrology: How Water Flows?

How does water move through the ground and interact with sediments and rock? Will an aquifer recharge slowly or quickly after water is withdrawn, and where will new groundwater come from? These questions are central for communities that need adequate drinking water, farmers tending crops and livestock, and engineers working to keep water supplies free of contaminants. For example, the 1986 trial recounted in the book and movie A Civil Action focused on town drinking wells in Woburn, Massachusetts, that were polluted with industrial chemicals suspected of causing cancer among residents. Plaintiffs asserted—and an investigation by the Environmental Protection Agency ultimately confirmed—that chemicals dumped by several local businesses had flowed through groundwater to the underlying aquifer and contaminated the wells.

The pore structure of soils, sediment, and rock is a central influence on groundwater movement. Hydrologists quantify this influence primarily in terms of:

- Porosity: The proportion of total volume that is occupied by voids, like the spaces within a pile of marbles. Porosity is not a direct function of the size of soil grains—the porosity of a pile of basketballs is the same as a pile of marbles. Porosity tends to be larger in well sorted sediments where the grain sizes are uniform, and smaller in mixed soils where smaller grains fill the voids between larger grains. Soils are less porous at deeper levels because the weight of overlying soil packs grains closer together.

- Permeability: How readily the medium transmits water, based on the size and shape of its pore spaces and how interconnected its pores are.

Materials with high porosity and high permeability, such as sand, gravel, sandstone, fractured rock, and basalt, produce good aquifers. Low-permeable rocks and sediments that impede groundwater flow include granite, shale, and clay.

Groundwater recharge enters aquifers in areas at higher elevations (typically hill slopes) than discharge areas (typically in the bottom of valleys), so the overall movement of groundwater is downhill. However, within an aquifer, water often flows upward toward a discharge area. To understand and map the complex patterns of groundwater flow, hydrogeologists use a quantity called the hydraulic head. The hydraulic head at a particular location within an aquifer is the sum of the elevation of that point and the height of the column of water that would fill a well open only at that point. Thus, the hydraulic head at a point is simply the elevation of water that rises up in a well open to the aquifer at that point.

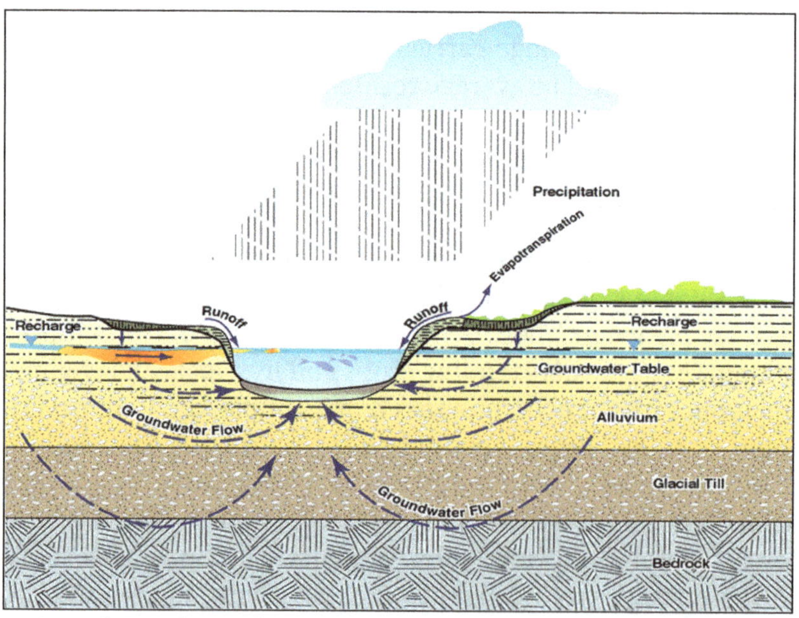

Groundwater flow under the Housatonic River, Pittsfield, Massachusetts.

The height of water within the well is not the same as the distance to the water table. If the aquifer is under pressure, or artesian, this height may be much greater than the distance to the water table. Thus the hydraulic head is the combination of two potentials: mechanical potential due to elevation, like a ball at the top of a ramp, and pressure potential, like air compressed in a balloon. Because these are usually the only two significant potentials driving groundwater flow, groundwater will flow from high to low hydraulic head.

This theory works in the same way that electrical potential (voltage) drives electrical flow and thermal potential (temperature) drives heat conduction. Like these other fluxes, groundwater flux between two points is simply proportional to the difference in potential, hydraulic head, and also to the permeability of the medium through which flow is taking place. These proportionalities are expressed in the fundamental equation for flow through porous media, known as Darcy's Law.

The gradient in hydraulic potential may drive groundwater flow downward, upward, or horizontally. Hydrogeologists collect water levels measured in wells to map hydraulic potential in aquifers. These maps can then be combined with permeability maps to determine the pattern in which groundwater flows throughout the aquifer.

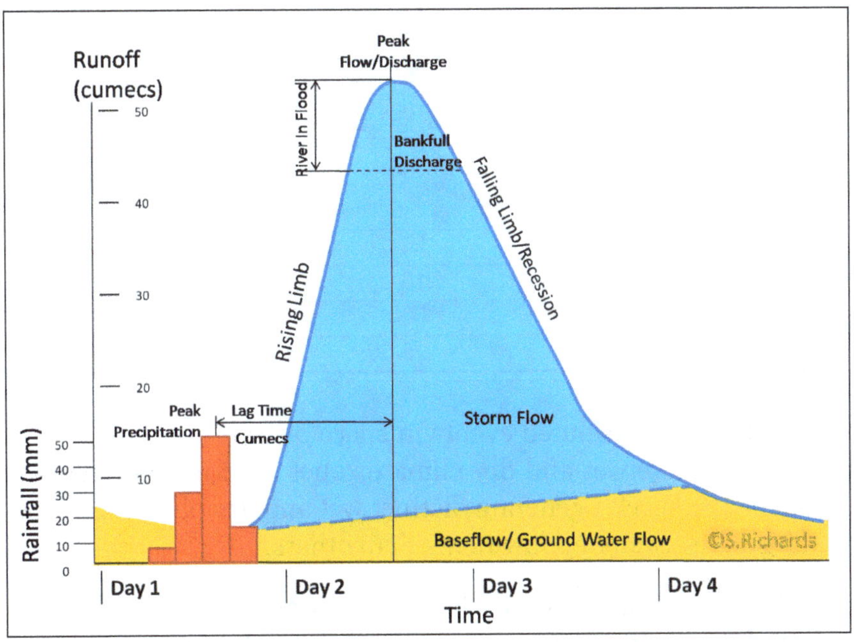

Components of a typical flood hydrograph.

Depending on local rainfall, land use, and geology, streams may be fed by either groundwater discharge or surface runoff and direct rainfall, or by some combination of surface and groundwater. Perennial streams and rivers are primarily supplied by groundwater, referred to as baseflow. During dry periods they are completely supplied by groundwater; during storms there is direct runoff and groundwater discharge also increases. The hydrograph shows flow patterns in a stream before, during, and after a storm with relative contributions from groundwater (baseflow) and surface water (quickflow, also referred to as storm flow).

World Demand for Water

How much water do humans use? The answer depends on where they live and on their socioeconomic status. Under primitive conditions a person will consume three to five gallons per day for drinking and subsistence farming. In a city where water is also used for cleaning, manufacturing, and sanitation, per capita use is around 150 gallons per day. In the United States, which has among the highest water consumption rates in the world, each person uses an average of 1,340 gallons of water per day. Table shows how much water is required to produce common goods and services.

Average water requirements.

Item	Gallons used
1 pound of cotton	2,000
1 pound of grain-fed beef	800
1 loaf of bread	150
1 car	100,000
1 kilowatt hour of electricity	25
1 pound of rubber	100

Item	Gallons used
1 pound of steel	25
1 gallon of gasoline	10
1 load of laundry	60
1 ten-minute shower	25-50

Water resources are not distributed evenly in space or time around the world. Global circulation patterns create wet and dry climate zones, and in some regions seasonal or multi-annual climate cycles generate distinct wet and dry phases. As a result, some regions have larger freshwater endowments than others.

Although developed nations generally have more water available than many countries in Africa and the Middle East, some areas with good water endowments still are subject to "water stress" because they are withdrawing water from available supplies at extremely high rates. High-intensity water uses in industrialized nations include agricultural production and electric power generation, which requires large quantities of water for cooling. In the United States electric power production accounts for 39 percent of all freshwater withdrawals, although almost all of this water is immediately returned to the rivers from which it is withdrawn. Agriculture consumes much more water because irrigation increases transpiration to the atmosphere.

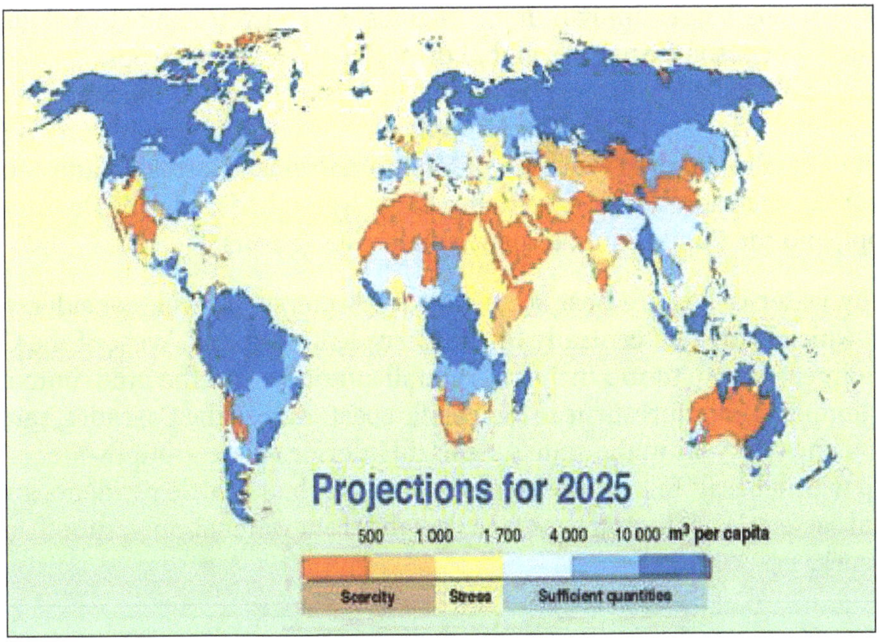

World freshwater supplies.

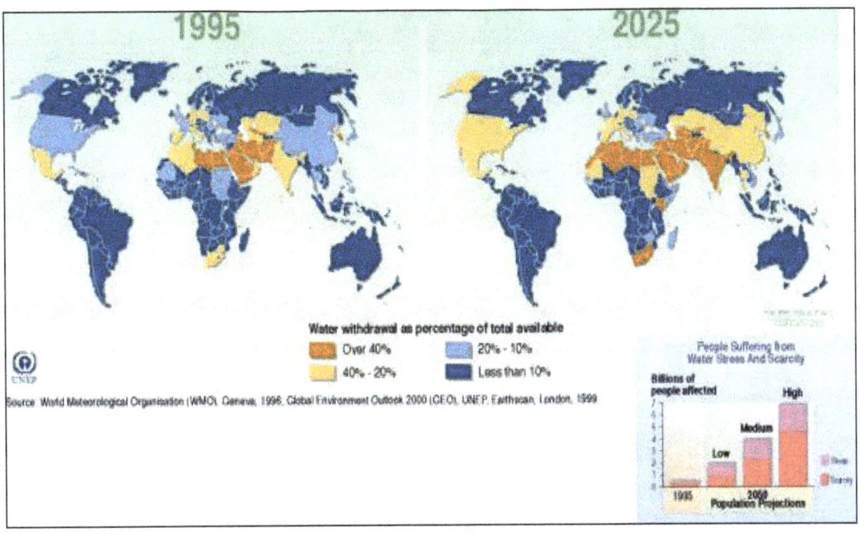

Current and projected freshwater stress areas.

As of 2002, 1.1 billion people around the world (17 percent of global population) did not have access to safe drinking water and 2.6 billion people (42 percent of global population) lived without adequate sanitation. As a result, millions of people die each year of preventable water-related diseases. Most of the countries with inadequate supplies of safe drinking water are located in Africa, Asia, and the Pacific, but problems persist elsewhere as well. For example, many households lack adequate sewage treatment services in Eastern Europe. And inequity among water users is widespread: cities often receive better service than rural areas, and many poor communities in both rural and urban areas lack clean water and sanitation.

Although these challenges apply in many regions, it is hard to make broad generalizations about water resources at the global or national level; to paraphrase the famous saying about politics, all hydrology is local. The basic geologic unit that scientists focus on to characterize an area's water supply and water quality with precision is the watershed or catchment area—an area of land that drains all streams and rainfall to a common outlet such as a bay or river delta. Large watersheds, such as the Amazon, the Mississippi, and the Congo contain many smaller sub-basins.

To see why water issues are best studied at the watershed level, consider Washington State, which is divided centrally by the Cascade Mountains. West of the Cascades, Washington receives up to 160 inches of rainfall annually, and the mild, humid climate supports temperate rainforests near the Pacific coast. Across the Cascades, rainfall is as low as six inches per year in the state's semiarid interior where groundwater is pumped from deep within basalt formations to grow wheat. Urban Seattle residents and ranchers in rural eastern Washington thus face very different water supply, runoff, and water quality issues.

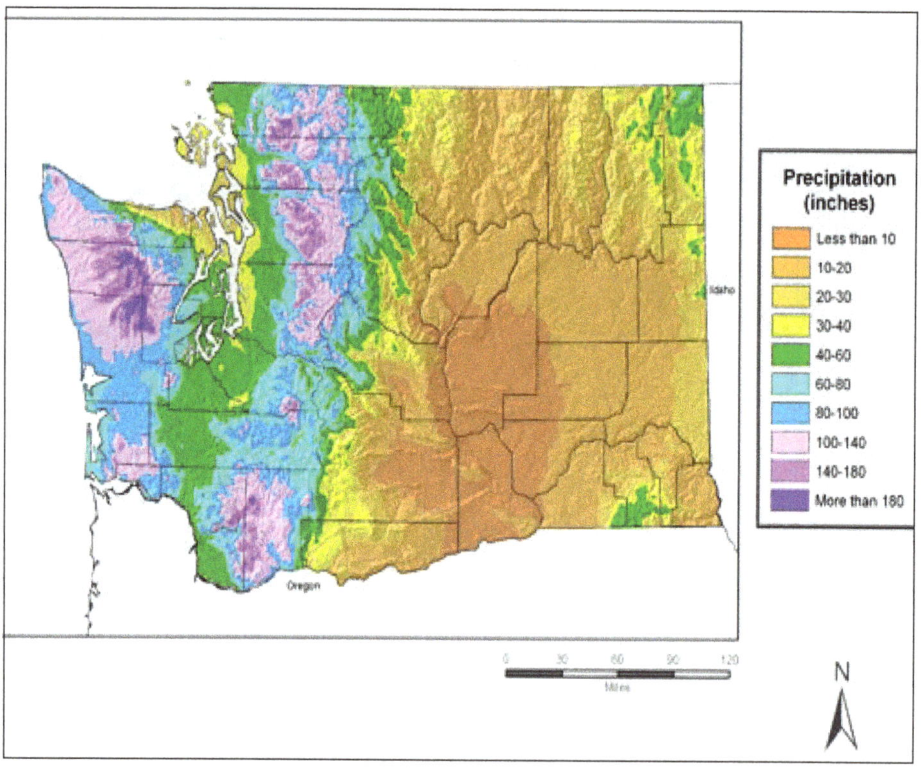

Average annual precipitation, Washington.

Currently 10,000 to 12,000 cubic kilometers of freshwater are available for human consumption each year worldwide. In the year 2000 humans withdrew about 4,000 km3 from this supply. About half of the water withdrawn was consumed, meaning that it was evaporated, transpired by plants, or contaminated beyond use, and so became temporarily unavailable for other users. The other 50 percent was returned to use: for example, some water used for irrigation drains back into rivers or recharges groundwater, and most urban wastewater is treated and returned to service.

Of the water withdrawn for human use, 65 percent went to agriculture, 10 percent to domestic use (households, municipal water systems, commercial use, and public services), 20 percent to industry (mostly electric power production), and 5 percent evaporated from reservoirs. About 70 percent of the water used for agriculture was consumed, compared to 14 percent of water used for domestic consumption and 11 percent of water used for industry.

Both population levels and economic development are important drivers of world water use. If current patterns continue, the World Water Council estimates that total yearly withdrawals will rise to more than 5,000 km3 by 2050 as world population rises from 6.1 billion to 9.2 billion. During the 20th century, world population tripled but water use rose by a factor of six. The United Nations and the international community have set goals of halving the number of people without adequate safe drinking water and sanitation by 2015. Meeting this target will require providing an additional 260,000 people per day with clean drinking water and an additional 370,000 people per day with improved sanitation through the year 2014, even as overall world demand for water is rising.

SURFACE WATER

Surface water includes water found in streams, rivers, lakes, marshland, snow, ocean water, or any other water found on Earth's surface. Groundwater is located in the subsurface in reservoirs (aquifers).

Approximately 70% of Earth's surface is covered by surface water. Of this estimated 1.4 billion cubic kilometers of water, almost 98% is saltwater and between 2 and 3% is freshwater. Surface water can be replenished by precipitation and can pass to the atmosphere via evaporation, and so is a crucial part of the planet's hydrologic cycle.

The increased temperature and drier climate that are developing in some regions of the world under the influence of global warming—the increasing warming of the atmosphere that is being driven in part by the production of compounds associated with human activity that restrict the dissipation of heat in the atmosphere—is causing some surface water sources to decrease or dry up altogether. In underdeveloped countries, this is a concern to human welfare and survival, where freshwater is scarce.

Surface water is also more susceptible to microbial contamination than is groundwater, and so is an important cause of waterborne illnesses such as cholera.

The surface water that is replenished by precipitation and which can release water vapor to the atmosphere is a small proportion of the total surface freshwater on Earth. Of the estimated 35 million cubic kilometers of fresh surface water, almost 25 million cubic kilometers are unavailable, as they are present as glacial ice, permafrost, or permanently frozen snow.

Evaporation of surface water adds an estimated 580,000 cubic kilometers of water vapor to the atmosphere every year. Without this infusion of water vapor, the planetary water cycle, and so life on Earth, would cease.

Because of its location, surface water is susceptible to contamination by a variety of pollutants and microorganisms. Typically, the disease-causing microbes normally live in the intestinal tract of warm-blooded animals including humans. They enter the surface water in feces. Examples of disease-causing bacteria are those in the genera of Salmonella, Shigella, Vibrio, and Escherichia.

Escherichia colio157:H7 has become prominent. Contamination of drinking water with 0157:H7 can be devastating. An infamous example was the contamination of the municipal water supply of Walkerton, Ontario, Canada, in 2000. Several thousand people became ill, and seven people died due to ingestion of the contaminated water that had flowed into an uncapped well from an adjacent cattle field.

Protozoa can also contaminate surface water. The main two examples are Giardia and Cryptosporidium, which are naturally present in the intestinal tract of beavers and deer. Waterborne illness due to these protozoans is increasing in North America, as human settlement encroaches on previously natural areas.

Impacts and Issues

The substantial proportion of fresh surface water that has been inaccessible is becoming more available as polar regions of Earth increase in temperature due to global warming. Large regions of glacial ice in the Antarctic have broken away into the surrounding sea. Eventually the ice will melt, increasing the volume of liquid water present in the global ocean.

Measurements of tide levels have been done throughout the twentieth century and satellite measurements have been ongoing since the 1970s. These measurements have established that the global sea level is rising. According to the 2007 report of the Intergovernmental Panel on Climate Change (IPCC), the average global sea level rose by 1.8 millimeters per year from 1961 to the early 1990s, and has been rising by over 3 millimeters per year since 1993.

Rising ocean waters are already affecting coastal regions. For example, according to the Geological Survey of Canada, rising sea levels are contributing to the increased damage of beach coastline in the maritime province of Prince Edward Island caused by coastal storms, which is threatening the province's nearly $3 million annual tourism industry.

If the rising surface waters continue, almost half the coastline could be submerged by 2100.

Florida is another area being affected. The advancing sea level has already affected surface water bodies including coastal wetlands. The low elevation of the state puts even the interior of Florida, including the Everglades, at risk, and so could threaten many species of birds and wildlife.

The fact that many of Florida's shorelines are gently sloping will increase the height of the sea-level rise, which could exceed 2 ft (0.6 m) by 2100. The state's drinking water supplies could be challenged.

In contrast, climate change is causing a decline in surface waters in regions of the globe that are becoming warmer and drier. For the 600 million people who already face water scarcities, the further loss of surface water could be life threatening. The World Health Organization estimates that by the year 2025, almost 3 billion people could be affected by inadequate water supplies.

LAKE

A lake is an area filled with water, localized in a basin, that is surrounded by land, apart from any river or other outlet that serves to feed or drain the lake. Lakes lie on land and are not part of the ocean, and therefore are distinct from lagoons, and are also larger and deeper than ponds, though there are no official or scientific definitions. Lakes can be contrasted with rivers or streams, which are usually flowing. Most lakes are fed and drained by rivers and streams.

Natural lakes are generally found in mountainous areas, rift zones, and areas with ongoing glaciation. Other lakes are found in endorheic basins or along the courses of mature rivers. In some parts of the world there are many lakes because of chaotic drainage patterns left over from the last Ice Age. All lakes are temporary over geologic time scales, as they will slowly fill in with sediments or spill out of the basin containing them.

Lake Sevan is the largest body of water in Armenia and the Caucasus region. It is one of the largest freshwater high-altitude (alpine) lakes in Eurasia.

Many lakes are artificial and are constructed for industrial or agricultural use, for hydro-electric power generation or domestic water supply, or for aesthetic, recreational purposes, or other activities.

Peyto Lake, Alberta, Canada Distribution of lakes.

The Seven Rila Lakes are a group of glacial lakes in the Bulgarian Rila mountains.

Wayanad District Kerala India.

The majority of lakes on Earth are freshwater, and most lie in the Northern Hemisphere at higher latitudes. Canada, with a deranged drainage system has an estimated 31,752 lakes larger than 3 square kilometres (1.2 sq mi) and an unknown total number of lakes, but is estimated to be at least 2 million. Finland has 187,888 lakes 500 square metres (5,400 sq ft) or larger, of which 56,000 are large (10,000 square metres (110,000 sq ft) or larger).

Most lakes have at least one natural outflow in the form of a river or stream, which maintain a lake's average level by allowing the drainage of excess water. Some lakes do not have a natural outflow and lose water solely by evaporation or underground seepage or both. They are termed endorheic lakes.

Many lakes are artificial and are constructed for hydro-electric power generation, aesthetic purposes, recreational purposes, industrial use, agricultural use or domestic water supply.

Evidence of extraterrestrial lakes exists; "definitive evidence of lakes filled with methane" was announced by NASA as returned by the Cassini Probe observing the moon Titan, which orbited the planet Saturn.

Globally, lakes are greatly outnumbered by ponds: of an estimated 304 million standing water bodies worldwide, 91% are 1 hectare (2.5 acres) or less in area. Small lakes are also much more numerous than large lakes: in terms of area, one-third of the world's standing water is represented by lakes and ponds of 10 hectares (25 acres) or less. However, large lakes account for much of the area of standing water with 122 large lakes of 1,000 square kilometres (390 sq mi, 100,000 ha, 247,000 acres) or more representing about 29% of the total global area of standing inland water.

Origin of Lakes

A portion of the Great Salt Lake in Utah, United States.

A lake in the Andes Mountains.

Tectonic Lakes

Tectonic lakes are lakes formed by the deformation and resulting lateral and vertical movements of the Earth's crust. These movements include faulting, tilting, folding, and warping. Some of the well-known and largest lakes on Earth are rift lakes occupying rift valleys, e.g., Central African Rift lakes and Lake Baikal. Other well-known tectonic lakes, Caspian Sea, the Sea of Aral, and other lakes from the Pontocaspian occupy basins that have been separated from the sea by the tectonic uplift of the sea floor above sea level.

Often, the tectonic action of crustal extension has created an alternating series of parallel grabens and horsts that form elongate basins alternating with mountain ranges. Not only does this promote the creation of lakes by the disruption of preexisting drainage networks, it also creates within arid regions endorheic basins that containing salt lakes (also called saline lakes). They form where there is no natural outlet, a high evaporation rate and the drainage surface of the water table has a higher-than-normal salt content. Examples of these salt lakes include Great Salt Lake and the Dead Sea. Another type of tectonic lake caused by faulting is sag ponds.

Volcanic Lakes

Volcanic lakes are lakes that occupy either local depressions, e.g., craters and maars or larger basins, e.g. calderas, created by volcanism. Crater lakes are formed in volcanic craters and calderas, which fill up with precipitation more rapidly than they empty via either evaporation, groundwater discharge, or combination of both. Sometimes the latter are called caldera lakes, although often no distinction is made. An example is Crater Lake in Oregon, in the caldera of Mount Mazama. The caldera was created in a massive volcanic eruption that led to the subsidence of Mount Mazama around 4860 BCE. Other volcanic lakes are created when either rivers or streams are dammed by lava flows or volcanic lahars. The basin which is now Malheur Lake, Oregon was created when a lava flow dammed the Malheur River.

Glacial Lakes

Glacial lakes are lakes created by the direct action of glaciers and continental ice sheets. A wide variety of glacial processes create enclosed basins. As a result, there are a wide variety of different types of glacial lakes and it is often difficult to define clear-cut distinctions between different types of glacial lakes and lakes influenced by other activities. The general types of glacial lakes that have recognized are lakes in direct contact with ice; glacially carved rock basins and depressions; morainic and outwash lakes; and glacial drift basins. Glacial lakes are the most numerous lakes in the world. Most lakes in northern Europe and North America have been either influenced or created by the latest, but not last, glaciation, to have covered the region. Glacial lakes include proglacial lakes, subglacial lakes, finger lakes, and epishelf lakes. Epishelf lakes are highly stratified lakes in which a layer of freshwater, derived from ice and snow melt, is dammed behind an ice shelf that is attached to the coastline. They are mostly found in Antarctica.

Fluvial Lakes

Fluvial (or riverine) lakes are lakes produced by running water. These lakes include plunge pool lakes, fluviatile dams and meander lakes.

Oxbow Lakes

The most common type of fluvial lake is a crescent-shaped lake called an oxbow lake due to the distinctive curved shape. They can form in river valleys as a result of meandering. The slow-moving river forms a sinuous shape as the outer side of bends are eroded away more rapidly than the inner side. Eventually a horseshoe bend is formed and the river cuts through the narrow neck. This new passage then forms the main passage for the river and the ends of the bend become silted up, thus forming a bow-shaped lake.

Fluviatile Dams

These form where sediment from a tributary blocks the main river.

Lateral Lakes

These form where sediment from the main river blocks a tributary, usually in the form of a levee.

Solution Lakes

A solution lake is a lake occupying a basin formed by surface dissolution of bedrock. In areas underlain by soluble bedrock, its solution by precipitation and percolating water commonly produce cavities. These cavities frequently collapse to form sinkholes that form part of the local karst topography. Where groundwater lies near the grounds surface, a sinkhole will be filled water as a solution lake. If such a lake consists of a large area of standing water that occupies an extensive closed depression in limestone, it is also called a karst lake. Smaller solution lakes that consist of a body of standing water in a closed depression within a karst region are known as karst ponds. Limestone caves often contain pools of standing water, which are known as underground lakes. Classic examples of solution lakes are abundant in the karst regions at the Dalmatian coast of Croatia and within large parts of Florida.

Landslide Lakes

Landslide lakes are lakes created by the blockage of a valley by either mudflows, rockslides, or screes. Such lakes are common in mountainous regions. Although landslide lakes may be large and quite deep, they are typically short-lived. An example of a landslide lake is Quake Lake, which formed as a result of the 1959 Hebgen Lake earthquake.

Aeolian Lakes

Aeolian lakes are lakes produced by wind action. They are found mainly in arid environments although some aeolian lakes are relict landforms indicative of arid paleoclimates. Aeolian lakes consist of lake basins dammed by wind-blown sand; interdunal lakes that lies between well-oriented sand dunes; and deflation basins formed by wind action under previously arid paleoenvironments. Moses Lake, Washington, is an example of a lake basins dammed by wind-blown sand.

Shoreline Lakes

Shoreline lakes are generally lakes created by blockage of estuaries or by the uneven accretion of beach ridges by longshore and other currents. They include maritime coastal lakes, ordinarily in drowned estuaries; lakes enclosed by two tombolos or spits connecting an island to the mainland; lakes cut off from larger lakes by a bar; or lakes divided by the meeting of two spits.

Organic Lakes

Organic lakes are lakes created by the actions of plants and animals. On the whole they are relatively rare in occurrence and quite small in size. In addition, they typically ephemeral features relative to the other types of lakes. The basins in which organic lakes occur are associated with beaver dams, coral lakes, or dams formed by vegetation.

Peat Lakes

Peat lakes are a form of organic lake. They form where a buildup of partly decomposed plant material in a wet environment leaves the vegetated surface below the water table for a sustained period of time. They are often low in nutrients and mildly acidic, with bottom waters low in dissolved oxygen.

Anthropogenic Lakes

Anthropogenic lakes are artificially created lakes formed by human activity. They can be the result of intentional damming of rivers and streams or subsequent filling of abandoned excavations by either ground water, precipitation, or a combination of both.

Meteorite (Extraterrestrial Impact/Crater) Lakes

Meteorite lakes, which are also known as crater lakes, are lakes created by catastrophic extraterrestrial impacts by either meteorites or asteroids. Examples of meteorite lakes are Lonar crater lake, India, Lake Elgygytgyn, and Pingualuit crater lake, Quebec, Canada, As in case of Lake El'gygytgyn and Pingualuit crater lake, meteorite (extraterrestrial impact/crater) lakes can contain unique and scientifically valuable sedimentary deposits associated with long records of paleoclimatic changes.

Other Different Types of Lakes

One of the many artificial lakes in Arizona at sunset.

The crater lake of Volcán Irazú,
Costa Rica.

These kettle lakes in Alaska were formed by a retreating glacier.

In addition to mode of origin, lakes have been named and classified in various other ways according to their thermal stratification, salinity, relative seasonal permanence, degree of outflow, and other factors. Also, different cultures and regional of the world have their popular nomenclature

Ice Melting on Lake Balaton.

Types of Lakes according to Thermal Stratification

In addition to their origin, there are various other ways of either naming or defining types of lakes. One major way of classification lakes in on the basis of thermal stratification because it is a major control on animal and plant life inhabiting a lake and the fate and distribution of dissolved and suspended material in a lake. For example, the thermal stratification and the degree and frequency of mixing exerts a strong control on the distribution of oxygen within it. In addition, lake can be classified according important factors such as seasonal variations in lake volume and level, oxygen saturation, and salinity of its water mass. Finally, the names of types of lakes that are used by the lay public and in the scientific for different types of lakes are often informally derived from either from their morphology of other aspects or their physical characteristics.

F.A. Forel, who is also referred to as the father of limnology, was the first scientist to classify lakes according to their thermal stratification. His system of classification was later modified and improved upon by Hutchinson and Laffler. Because the density of water varies with temperature, with a maximum at +4 DC, thermal stratification is an important physical characteristic of lakes that controls the fauna and flora, sedimentation, chemistry, and other aspects of individual lakes. First, the colder, heavier water typically forms a layer near the bottom, which called the hypolimnion. Second, normally overlying it is a transition zone known as the metalimnion. Finally, overlying the metalimnion is a surface layer of a warmer, lighter water is called the epilimnion. However, this typical stratification sequence can vary widely depending either on the specific lake, the time of season, or combination of both.

Based upon thermal stratification, lakes are classified as either holomictic lakes or meromictic lakes. A meromictic lake is a lake which has layers of water which do not intermix. The deepest layer of water in such a lake does not contain any dissolved oxygen. In addition, the layers of sediment at the bottom of a meromictic lake remain relatively undisturbed because there are no living aerobic organisms. The lack of disturbance allows for the development of lacustrine varves. A Holomictic lake is a lake that has a uniform temperature and density from top to bottom at a specific time during the year. This uniformity temperature and density in allows the lake waters to completely mix. Holomictic lakes are non-meromictic lakes. Based upon thermal stratification and frequency of turnover, holomictic lakes are divided into amictic lakes, cold monomictic lakes, dimictic lakes, warm monomictic lakes, polymictic lakes, and oligomictic lakes. The classification of lakes by thermal stratification presupposes lakes with sufficient depth to form a hypolimnion. As a results, very shallow lakes are excluded this classification system.

The stratification in a lake is not always the result of variation to density because of thermal gradients. Stratification within a lake can also be the result of differences in density resulting from gradients in salinity. In case of a difference in salinity, the hypolimnion and epilimnion are separated not by a thermocline but by a halocline, which is sometimes referred to as a chemocline.

Types of Lake according to Seasonal Variation of Lake Level and Volume

Lakes are informally classified and named according to the seasonal variation in their lake level and volume. Some of the names include:

- Ephemeral lake is a short-lived lake or pond. If it fills with water and dries up (disappears) seasonally it is known as an intermittent lake They often fill poljes.

- Dry lake is a popular name for an ephemeral lake that contains water only intermediately at irregular and infrequent intervals.
- Perennial lake is a lake that has water in its basin throughout the year and is not subject to extreme fluctuations in level.
- Playa lake is a typically shallow, intermittent lake that covers or occupies a playa either in wet seasons or in especially wet years but subsequently drying up in an arid or semiarid region.
- Vlei is a name used in South Africa for a shallow lake which varies considerably in level with the seasons.

Types of Lake according to Water Chemistry

Lakes are also informally classified and named according to the general chemistry of their water mass. Some of the types of lakes include:

- An acid lake is a lake that has a pH is below neutral (<6.5). A lake is considered highly acidic when the pH drops below 5.5, below which when biological consequences occur. Such lakes include acid pit lakes occupying abandoned mines and excavations; naturally acid lakes of igneous and metamorphic landscapes; peat bogs in northern regions; acid-saline lakes of arid environments; crater lakes of active and dormant volcanoes; and lakes acidified by acid rain.
- A salt lake, which also known as a *brine lake*, is an inland body of water situated in an arid or semiarid region, having no outlet to the sea, and containing a high concentration of dissolved salts (principally sodium chloride). Examples include the Great Salt Lake in Utah, and the Dead Sea in the Near East.
- Alkali sink, also known as salt flats, are lakes on the other extreme of the scale from the large and deep saline lakes. They are, shallow saline features that occupy low-lying areas of the arid regions and in groundwater discharge zones. These are typically classifiable as either playas or playa lakes because they are periodically flooded by either rain or flood events and then dry up during drier intervals, leaving accumulations of brines and evaporitic minerals.
- A salt pan (saltpan) is either an undrained small shallow natural depression in which water accumulates and evaporates leaving a salt deposit or the shallow lake of brackish water occupying a salt pan. This term is also used for a large pan for recovering salt by evaporation.
- A saline pan is a name for an ephemeral saline lake which precipitates a bottom crust that is subsequently modified during subaerial exposure.

Lakes Composed of other Liquids

- Lava lake is a large volume of molten lava, usually basaltic, contained in a volcanic vent, crater, or broad depression.
- Hydrocarbon lakes are bodies of liquid ethane and methane that occupy depressions on the surface of Titan. They were detected by the Cassini–Huygens space probe.

Paleolakes

A paleolake, also spelt palaeolake, is a lake that existed in the past when hydrological conditions were different. Quaternary paleolakes can often be identified on the basis of relict lacustrine landforms such as relict lake plains and coastal landforms that form recognizable relict shorelines, which are called paleoshorelines. Paleolakes can also be recognized by characteristic sedimentary deposits that accumulated in them and any fossils that these sediments might contain. The paleoshorelines and sedimentary deposits of paleolakes provide evidence for prehistoric hydrological changes during the times that they existed.

Types of paleolakes include:

- A former lake is a lake which is no longer in existence. Such lakes include prehistoric lakes and lakes which have permanently dried up often as the result of either evaporation or human intervention. Owens Lake in California, USA, is an example of a former lake. Former lakes are a common feature of the Basin and Range area of southwestern North America.
- A shrunken lake is a lake which has drastically decreased in size over geological time. Lake Agassiz, which once covered much of central North America, is a good example of a shrunken lake. Two notable remnants of this lake are Lake Winnipeg and Lake Winnipegosis.

Paleolakes are of scientific and economic importance. For example, Quaternary paleolakes in semidesert basins are important for two reasons. First, they played an extremely significant, if transient, role in shaping the floors and piedmonts of many basins. Finally, their sediments contain enormous quantities of geologic and paleontologic information concerning past environments. In addition, the organic-rich deposits of pre-Quaternary paleolakes are important either for the thick deposits of oil shale and shale gas that they contain or as source rocks of petroleum and natural gas. Although of significantly less economic importance, strata deposited along the shore of paleolakes sometimes contain coal seams.

Characteristics

Numerous features in addition to lake type, such as drainage basin (also known as catchment area), inflow and outflow, nutrient content, dissolved oxygen, pollutants, pH, and sedimentation.

Many lakes can have tremendous cultural importance. The West Lake of Hangzhou has inspired romantic poets throughout the ages, and has been an important influence on garden designs in China, Japan and Korea.

Changes in the level of a lake are controlled by the difference between the input and output compared to the total volume of the lake. Significant input sources are precipitation onto the lake, runoff carried by streams and channels from the lake's catchment area, groundwater channels and aquifers, and artificial sources from outside the catchment area. Output sources are evaporation from the lake, surface and groundwater flows, and any extraction of lake water by humans. As climate conditions and human water requirements vary, these will create fluctuations in the lake level.

Lake Päijänne is one of tens of thousands of lakes in Finnish Lakeland.

Lake Mapourika, New Zealand Lakes.

Lakes can be also categorized on the basis of their richness in nutrients, which typically affect plant growth. Nutrient-poor lakes are said to be oligotrophic and are generally clear, having a low concentration of plant life. Mesotrophic lakes have good clarity and an average level of nutrients. Eutrophic lakes are enriched with nutrients, resulting in good plant growth and possible algal blooms. Hypertrophic lakes are bodies of water that have been excessively enriched with nutrients. These lakes typically have poor clarity and are subject to devastating algal blooms. Lakes typically reach this condition due to human activities, such as heavy use of fertilizers in the lake catchment area. Such lakes are of little use to humans and have a poor ecosystem due to decreased dissolved oxygen.

Due to the unusual relationship between water's temperature and its density, lakes form layers called thermoclines, layers of drastically varying temperature relative to depth. Fresh water is most dense at about 4 degrees Celsius (39.2 °F) at sea level. When the temperature of the water at the surface of a lake reaches the same temperature as deeper water, as it does during the cooler months in temperate climates, the water in the lake can mix, bringing oxygen-starved water up from the depths and bringing oxygen down to decomposing sediments. Deep temperate lakes can maintain a reservoir of cold water year-round, which allows some cities to tap that reservoir for deep lake water cooling.

Lake Teletskoye, Siberia.

Since the surface water of deep tropical lakes never reaches the temperature of maximum density, there is no process that makes the water mix. The deeper layer becomes oxygen starved and can become saturated with carbon dioxide, or other gases such as sulfur dioxide if there is even a trace of volcanic activity. Exceptional events, such as earthquakes or landslides, can cause mixing which rapidly brings the deep layers up to the surface and release a vast cloud of gas which lay trapped in solution in the colder water at the bottom of the lake. This is called a limnic eruption. An example is the disaster at Lake Nyos in Cameroon. The amount of gas that can be dissolved in water is directly related to pressure. As deep water surfaces, the pressure drops and a vast amount of gas comes out of solution. Under these circumstances carbon dioxide is hazardous because it is heavier than air and displaces it, so it may flow down a river valley to human settlements and cause mass asphyxiation.

The material at the bottom of a lake, or lake bed, may be composed of a wide variety of inorganics, such as silt or sand, and organic material, such as decaying plant or animal matter. The composition of the lake bed has a significant impact on the flora and fauna found within the lake's environs by contributing to the amounts and the types of nutrients available.

A paired (black and white) layer of the varved lake sediments correspond to a year. During winter, when organisms die, carbon is deposited down, resulting to a black layer. At the same year, during summer, only few organic materials are deposited, resulting to a white layer at the lake bed. These are commonly used to track past paleontological events.

Natural lakes provide a microcosm of living and nonliving elements that are relatively independent of their surrounding environments. Therefore, lake organisms can often be studied in isolation from the lake's surroundings.

Since the surface water of deep tropical lakes never reaches the temperature of maximum density, there is no process that makes the water mix. The deeper layer becomes oxygen starved and can become saturated with carbon dioxide, or other gases such as sulfur dioxide if there is even a trace of volcanic activity. Exceptional events, such as earthquakes or landslides, can cause mixing which rapidly brings the deep layers up to the surface and release a vast cloud of gas which lay trapped in solution in the colder water at the bottom of the lake. This is called a limnic eruption. An example is the disaster at Lake Nyos in Cameroon. The amount of gas that can be dissolved in water is directly related to pressure. As deep water surfaces, the pressure drops and a vast amount of gas comes out of solution. Under these circumstances carbon dioxide is hazardous because it is heavier than air and displaces it, so it may flow down a river valley to human settlements and cause mass asphyxiation.

The material at the bottom of a lake, or lake bed, may be composed of a wide variety of inorganics, such as silt or sand, and organic material, such as decaying plant or animal matter. The composition of the lake bed has a significant impact on the flora and fauna found within the lake's environs by contributing to the amounts and the types of nutrients available.

A paired (black and white) layer of the varved lake sediments correspond to a year. During winter, when organisms die, carbon is deposited down, resulting to a black layer. At the same year, during summer, only few organic materials are deposited, resulting to a white layer at the lake bed. These are commonly used to track past paleontological events.

Natural lakes provide a microcosm of living and nonliving elements that are relatively independent of their surrounding environments. Therefore, lake organisms can often be studied in isolation from the lake's surroundings.

Limnology

Lake Cügün, Kırşehir, Turkey.

Lake of Flowers (Liqeni i Lulëve), one of the Lurë Mountains glacial lakes, Albania.

Limnology is the study of inland bodies of water and related ecosystems. Limnology divides lakes into three zones: the littoral zone, a sloped area close to land; the photic or open-water zone, where sunlight is abundant; and the deep-water profundal or benthic zone, where little sunlight can reach. The depth to which light can reach in lakes depends on turbidity, determined by the density and size of suspended particles. A particle is in suspension if its weight is less than the random turbidity forces acting upon it. These particles can be sedimentary or biological in origin and are responsible for the color of the water. Decaying plant matter, for instance, may be responsible for a yellow or brown color, while algae may cause greenish water. In very shallow water bodies, iron oxides make water reddish brown. Biological particles include algae and detritus. Bottom-dwelling detritivorous fish can be responsible for turbid waters, because they stir the mud in search of food. Piscivorous fish contribute to turbidity by eating plant-eating (planktonivorous) fish, thus increasing the amount of algae. The light depth or transparency is measured by using a Secchi disk, a 20-cm (8 in) disk with alternating white and black quadrants. The depth at which the disk is no longer visible is the Secchi depth, a measure of transparency. The Secchi disk is commonly used to test for eutrophication. For a detailed look at these processes.

Lake Billy Chinook, Deschutes National Forest, Oregon.

A lake moderates the surrounding region's temperature and climate because water has a very high specific heat capacity (4,186 $J \cdot kg^{-1} \cdot K^{-1}$). In the daytime a lake can cool the land beside it with local winds, resulting in a sea breeze; in the night it can warm it with a land breeze.

How Lakes Disappear?

Lake Chad in a 2001 satellite image, with the actual lake in blue, and vegetation on top of the old lake bed in green.

Lake Badwater. Landsat 5 satellite photo.

Badwater Basin dry lake. Landsat 5 satellite photo.

The lake may be infilled with deposited sediment and gradually become a wetland such as a swamp or marsh. Large water plants, typically reeds, accelerate this closing process significantly because they partially decompose to form peat soils that fill the shallows. Conversely, peat soils in a marsh can naturally burn and reverse this process to recreate a shallow lake resulting in a dynamic equilibrium between marsh and lake. This is significant since wildfire has been largely suppressed in the developed world over the past century. This has artificially converted many shallow lakes into emergent marshes. Turbid lakes and lakes with many plant-eating fish tend to disappear more slowly. A "disappearing" lake (barely noticeable on a human timescale) typically has extensive plant mats at the water's edge. These become a new habitat for other plants, like peat moss when conditions are right, and animals, many of which are very rare. Gradually the lake closes and young peat may form, forming a fen. In lowland river valleys where a river can meander, the presence of peat is explained by the infilling of historical oxbow lakes. In the very last stages of succession, trees can grow in, eventually turning the wetland into a forest.

Some lakes can disappear seasonally. These are called intermittent lakes, ephemeral lakes, or seasonal lakes and can be found in karstic terrain. A prime example of an intermittent lake is Lake Cerknica in Slovenia or Lag Prau Pulte in Graubünden. Other intermittent lakes are only the result of above-average precipitation in a closed, or endorheic basin, usually filling dry lake beds. This can occur in some of the driest places on earth, like Death Valley. This occurred in the spring of 2005, after unusually heavy rains. The lake did not last into the summer, and was quickly evaporated. A more commonly filled lake of this type is Sevier Lake of west-central Utah.

Sometimes a lake will disappear quickly. On 3 June 2005, in Nizhny Novgorod Oblast, Russia, a lake called Lake Beloye vanished in a matter of minutes. News sources reported that government officials theorized that this strange phenomenon may have been caused by a shift in the soil underneath the lake that allowed its water to drain through channels leading to the Oka River.

The presence of ground permafrost is important to the persistence of some lakes. Thawing permafrost may explain the shrinking or disappearance of hundreds of large Arctic lakes across western Siberia. The idea here is that rising air and soil temperatures thaw permafrost, allowing the lakes to drain away into the ground.

Some lakes disappear because of human development factors. The shrinking Aral Sea is described as being "murdered" by the diversion for irrigation of the rivers feeding it.

Extraterrestrial Lakes

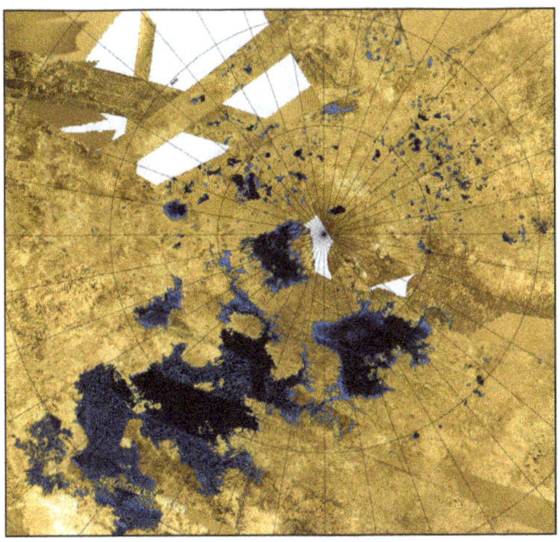

Titan's north polar hydrocarbon seas and lakes as seen in a false-color Cassini synthetic aperture radar mosaic.

Only one world other than Earth is known to harbor large lakes, Saturn's largest moon, Titan. Photographs and spectroscopic analysis by the Cassini–Huygens spacecraft show liquid ethane on the surface, which is thought to be mixed with liquid methane. The largest Titanean lake, Kraken Mare at 400,000 km2, is three-times the size of any lake on Earth, and even the second, Ligeia Mare, is estimated to be slightly larger than Earth's Lake Michigan–Huron.

Jupiter's large moon Io is volcanically active, and as a result sulfur deposits have accumulated on the surface. Some photographs taken during the Galileo mission appear to show lakes of liquid sulfur in volcanic caldera, though these are more analogous to lake of lava than of water on Earth.

The planet Mars has only one confirmed lake; it is underground and near the south pole. However, the surface of Mars is too cold and has too little atmospheric pressure to permit permanent surface water. Geologic evidence appears to confirm, however, that ancient lakes once formed on the surface. It is also possible that volcanic activity on Mars will occasionally melt subsurface ice, creating large temporary lakes. This water would quickly freeze and then sublimate, unless insulated in some manner, such as by a coating of volcanic ash.

There are dark basaltic plains on the Moon, similar to lunar maria but smaller, that are called lacus because they were thought by early astronomers to be lakes of water.

RIVER

Samur River in Azerbaijan – In the natural landscape.

Daugava river, Riga in Latvia – In the urban landscape.

A river is a natural flowing watercourse, usually freshwater, flowing towards an ocean, sea, lake or another river. In some cases a river flows into the ground and becomes dry at the end of its course without reaching another body of water. Small rivers can be referred to using names such as stream, creek, brook, rivulet, and rill. There are no official definitions for the generic term river as applied to geographic features, although in some countries or communities a stream is defined by its size. Many names for small rivers are specific to geographic location; examples are "run" in some parts of the United States, "burn" in Scotland and northeast England, and "beck" in northern England. Sometimes a river is defined as being larger than a creek, but not always: the language is vague.

Rivers are part of the hydrological cycle; water generally collects in a river from precipitation through a drainage basin from surface runoff and other sources such as groundwater recharge, springs, and the release of stored water in natural ice and snowpacks (e.g., from glaciers). Potamology is the scientific study of rivers, while limnology is the study of inland waters in general. Most of the major cities of the world are situated on the banks of rivers, as they are, or were, used as a source of water, for obtaining food, for transport, as borders, as a defensive measure, as a source of hydropower to drive machinery, for bathing, and as a means of disposing of waste.

Topography

Melting toe of Athabasca Glacier, Jasper National Park, Alberta, Canada.

The Loboc River in Bohol, Philippines.

A river begins at a source (or more often several sources), follows a path called a course, and ends at a mouth or mouths. The water in a river is usually confined to a channel, made up of a stream bed between banks. In larger rivers there is often also a wider floodplain shaped by flood-waters over-topping the channel. Floodplains may be very wide in relation to the size of the river channel. This distinction between river channel and floodplain can be blurred, especially in urban areas where the floodplain of a river channel can become greatly developed by housing and industry.

The Colorado River at Horseshoe Bend, Arizona.

Rivers can flow down mountains, through valleys (depressions) or along plains, and can create canyons or gorges.

The term upriver (or upstream) refers to the direction towards the source of the river, i.e. against the direction of flow. Likewise, the term downriver (or downstream) describes the direction towards the mouth of the river, in which the current flows.

The term left bank refers to the left bank in the direction of flow, right bank to the right.

The river channel typically contains a single stream of water, but some rivers flow as several interconnecting streams of water, producing a braided river. Extensive braided rivers are now found in only a few regions worldwide, such as the South Island of New Zealand. They also occur on peneplains and some of the larger river deltas. Anastamosing rivers are similar to braided rivers and are quite rare. They have multiple sinuous channels carrying large volumes of sediment. There are rare cases of river bifurcation in which a river divides and the resultant flows ending in different seas. An example is the bifurcation of Nerodime River in Kosovo.

The River Cam from the Green Dragon Bridge, Cambridge (Britain).

A river flowing in its channel is a source of energy which acts on the river channel to change its shape and form. In 1757, the German hydrologist Albert Brahms empirically observed that the submerged weight of objects that may be carried away by a river is proportional to the sixth power of the river flow speed. This formulation is also sometimes called Airy's law. Thus, if the speed of flow is doubled, the flow would dislodge objects with 64 times as much submerged weight. In mountainous torrential zones this can be seen as erosion channels through hard rocks and the creation of sands and gravels from the destruction of larger rocks. A river valley that was created from a U-shaped glaciated valley, can often easily be identified by the V-shaped channel that it has carved. In the middle reaches where a river flows over flatter land, meanders may form through erosion of the river banks and deposition on the inside of bends. Sometimes the river will cut off a loop, shortening the channel and forming an oxbow lake or billabong. Rivers that carry large amounts of sediment may develop conspicuous deltas at their mouths. Rivers whose mouths are in saline tidal waters may form estuaries.

Throughout the course of the river, the total volume of water transported downstream will often be a combination of the free water flow together with a substantial volume flowing through sub-surface rocks and gravels that underlie the river and its floodplain (called the hyporheic zone). For many rivers in large valleys, this unseen component of flow may greatly exceed the visible flow.

Subsurface Streams

Most but not all rivers flow on the surface. Subterranean rivers flow underground in caves or caverns. Such rivers are frequently found in regions with limestone geologic formations. Subglacial streams are the braided rivers that flow at the beds of glaciers and ice sheets, permitting meltwater to be discharged at the front of the glacier. Because of the gradient in pressure due to the overlying weight of the glacier, such streams can even flow uphill.

Permanence of Flow

An intermittent river (or ephemeral river) only flows occasionally and can be dry for several years at a time. These rivers are found in regions with limited or highly variable rainfall, or can occur because of geologic conditions such as a highly permeable river bed. Some ephemeral rivers flow during the summer months but not in the winter. Such rivers are typically fed from chalk aquifers which recharge from winter rainfall. In England these rivers are called bournes and give their name to places such as Bournemouth and Eastbourne. Even in humid regions, the location where flow begins in the smallest tributary streams generally moves upstream in response to precipitation and downstream in its absence or when active summer vegetation diverts water for evapotranspiration. Normally-dry rivers in arid zones are often identified as arroyos or other regional names.

The meltwater from large hailstorms can create a slurry of water, hail and sand or soil, forming temporary rivers.

Classification

Nile River delta, as seen from Earth orbit. The Nile is an example of a wave-dominated delta that has the classic Greek letter delta (Δ) shape after which river deltas were named.

A radar image of a 400-kilometre (250 mi) river of methane and ethane near the north pole of Saturn's moon Titan.

Rivers have been classified by many criteria including their topography, their biotic status, and their relevance to white water rafting or canoeing activities.

Topographical Classification

Rivers can generally be classified as either alluvial, bedrock, or some mix of the two. Alluvial rivers have channels and floodplains that are self-formed in unconsolidated or weakly consolidated sediments. They erode their banks and deposit material on bars and their floodplains. Bedrock rivers form when the river downcuts through the modern sediments and into the underlying bedrock. This occurs in regions that have experienced some kind of uplift (thereby steepening river gradients) or in which a particular hard lithology causes a river to have a steepened reach that has not been covered in modern alluvium. Bedrock rivers very often contain alluvium on their beds; this material is important in eroding and sculpting the channel. Rivers that go through patches of bedrock and patches of deep alluvial cover are classified as mixed bedrock-alluvial.

Alluvial rivers can be further classified by their channel pattern as meandering, braided, wandering, anastomose, or straight. The morphology of an alluvial river reach is controlled by a combination of sediment supply, substrate composition, discharge, vegetation, and bed aggradation.

At the start of the 20th century William Morris Davis devised the "cycle of erosion" method of classifying rivers based on their "age". Although Davis's system is still found in many books today, after the 1950s and 1960s it became increasingly criticized and rejected by geomorphologists. His scheme did not produce testable hypotheses and was therefore deemed non-scientific. Examples of Davis's river "ages" include:

- Youthful river: A river with a steep gradient that has very few tributaries and flows quickly. Its channels erode deeper rather than wider. Examples are the Brazos, Trinity and Ebro rivers.
- Mature river: A river with a gradient that is less steep than those of youthful rivers and flows more slowly. A mature river is fed by many tributaries and has more discharge than a youthful river. Its channels erode wider rather than deeper. Examples are the Mississippi, Saint Lawrence, Danube, Ohio, Thames and Paraná rivers.
- Old river: A river with a low gradient and low erosive energy. Old rivers are characterized by flood plains. Examples are the Yellow, lower Ganges, Tigris, Euphrates, Indus and lower Nile rivers.
- Rejuvenated river: A river with a gradient that is raised by tectonic uplift. Examples are the Rio Grande and Colorado River.

The ways in which a river's characteristics vary between its upper and lower course are summarized by the Bradshaw model. Power-law relationships between channel slope, depth, and width are given as a function of discharge by "river regime".

Biotic Classification

There are several systems of classification based on biotic conditions typically assigning classes from the most oligotrophic or unpolluted through to the most eutrophic or polluted. Other systems are based on a whole eco-system approach such as developed by the New Zealand Ministry for the Environment. In Europe, the requirements of the Water Framework Directive has led to the development of a wide range of classification methods including classifications based on fishery status A system of river zonation used in francophone communities divides rivers into three primary zones:

- The *crenon* is the uppermost zone at the source of the river. It is further divided into the eucrenon (spring or boil zone) and the hypocrenon (brook or headstream zone). These areas have low temperatures, reduced oxygen content and slow moving water.
- The *rhithron* is the upstream portion of the river that follows the crenon. It has relatively cool temperatures, high oxygen levels, and fast, turbulent, swift flow.
- The *potamon* is the remaining downstream stretch of river. It has warmer temperatures, lower oxygen levels, slow flow and sandier bottoms.

Whitewater Classification

The International Scale of River Difficulty is used to rate the challenges of navigation—particularly those with rapids. Class I is the easiest and Class VI is the hardest.

Stream Order Classification

The Strahler Stream Order ranks rivers based on the connectivity and hierarchy of contributing tributaries. Headwaters are first order while the Amazon River is twelfth order. Approximately 80% of the rivers and streams in the world are of the first and second order.

In certain languages, distinctions are made among rivers based on their stream order. In French, for example, rivers that run to the sea are called fleuve, while other rivers are called rivière. For example, in Canada, the Churchill River in Manitoba is called la rivière Churchill as it runs to Hudson Bay, but the Churchill River in Labrador is called le fleuve Churchill as it runs to the Atlantic Ocean. As most rivers in France are known by their names only without the word rivière or fleuve (e.g. la Seine, not le fleuve Seine, even though the Seine is classed as a fleuve), one of the most prominent rivers in the Francophonie commonly known as fleuve is le fleuve Saint-Laurent (the Saint Lawrence River).

Since many fleuves are large and prominent, receiving many tributaries, the word is sometimes used to refer to certain large rivers that flow into other fleuves; however, even small streams that run to the sea are called fleuve (e.g. fleuve côtier, "coastal fleuve").

Uses

Leisure activities on the River Avon at Avon Valley Country Park, Keynsham, United Kingdom. A boat giving trips to the public passes a moored private boat.

Rivers have been a source of food since pre-history. They are often a rich source of fish and other edible aquatic life, and are a major source of fresh water, which can be used for drinking and irrigation. Rivers help to determine the urban form of cities and neighbourhoods and their corridors often present opportunities for urban renewal through the development of foreshoreways such as river walks. Rivers also provide an easy means of disposing of waste water and, in much of the less developed world, other wastes.

Rivers have been used for navigation for thousands of years. The earliest evidence of navigation is found in the Indus Valley Civilization, which existed in northwestern India around 3300 BC. Riverine navigation provides a cheap means of transport, and is still used extensively on most major rivers of the world like the Amazon, the Ganges, the Nile, the Mississippi, and the Indus. Since river boats are often not regulated, they contribute a large amount to global greenhouse gas emissions, and to local cancer due to inhaling of particulates emitted by the transports.

Rivers have been important in determining political boundaries and defending countries. For example, the Danube was a long-standing border of the Roman Empire, and today it forms most of the border between Bulgaria and Romania. The Mississippi in North America and the Rhine in Europe are major east-west boundaries in those continents. The Orange and Limpopo Rivers in southern Africa form the boundaries between provinces and countries along their routes.

In some heavily forested regions such as Scandinavia and Canada, lumberjacks use the river to float felled trees downstream to lumber camps for further processing, saving much effort and cost by transporting the huge heavy logs by natural means.

Watermill in Belgium.

Fast flowing rivers and waterfalls are widely used as sources of energy, via watermills and hydroelectric plants. Evidence of watermills shows them in use for many hundreds of years, for instance in Orkney at Dounby Click Mill. Prior to the invention of steam power, watermills for grinding cereals and for processing wool and other textiles were common across Europe. In the 1890s the first machines to generate power from river water were established at places such as Cragside in Northumberland and in recent decades there has been a significant increase in the development of large scale power generation from water, especially in wet mountainous regions such as Norway.

The coarse sediments, gravel, and sand, generated and moved by rivers are extensively used in construction. In parts of the world this can generate extensive new lake habitats as gravel pits re-fill with water. In other circumstances it can destabilise the river bed and the course of the river and cause severe damage to spawning fish populations which rely on stable gravel formations for egg laying. In upland rivers, rapids with whitewater or even waterfalls occur. Rapids are often used for recreation, such as whitewater kayaking.

Ecosystem

The organisms in the riparian zone respond to changes in river channel location and patterns of flow. The ecosystem of rivers is generally described by the river continuum concept, which has some additions and refinements to allow for dams and waterfalls and temporary extensive flooding. The concept describes the river as a system in which the physical parameters, the availability of food particles and the composition of the ecosystem are continuously changing along its length. The food (energy) that remains from the upstream part is used downstream.

The general pattern is that the first order streams contain particulate matter (decaying leaves from the surrounding forests) which is processed there by shredders like Plecoptera larvae. The products of these shredders are used by collectors, such as Hydropsychidae, and further downstream algae that create the primary production become the main food source of the organisms. All changes are gradual and the distribution of each species can be described as a normal curve, with the highest density where the conditions are optimal. In rivers succession is virtually absent and the composition of the ecosystem stays fixed in time.

Chemistry

The chemistry of rivers is complex and depends on inputs from the atmosphere, the geology through which it travels and the inputs from man's activities. The chemical composition of the water has a large impact on the ecology of that water for both plants and animals and it also affects the uses that may be made of the river water. Understanding and characterising river water chemistry requires a well designed and managed sampling and analysis.

Brackish Water

Some rivers generate brackish water by having their river mouth in the ocean. This, in effect creates a unique environment in which certain species are found.

Flooding

Flooding is a natural part of a river's cycle. The majority of the erosion of river channels and the erosion and deposition on the associated floodplains occur during the flood stage. In many developed areas, human activity has changed the form of river channels, altering magnitudes and frequencies of flooding. Some examples of this are the building of levees, the straightening of channels, and the draining of natural wetlands. In many cases human activities in rivers and floodplains have dramatically increased the risk of flooding. Straightening rivers allows water to flow more rapidly downstream, increasing the risk of flooding places further downstream. Building on flood plains removes flood storage, which again exacerbates downstream flooding. The building of levees only protects the area behind the levees and not those further downstream. Levees and flood-banks can also increase flooding upstream because of the back-water pressure as the river flow is impeded by the narrow channel banks.

Flash flooding caused by a large amount of rain falling in a short amount of time.

Flow

Studying the flows of rivers is one aspect of hydrology.

Direction

Rivers flow downhill with their power derived from gravity. The direction can involve all directions of the compass and can be a complex meandering path.

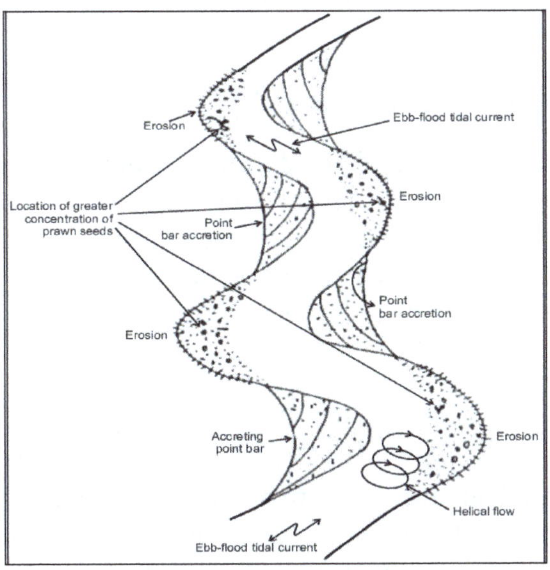

River meandering course.

Rivers flowing downhill, from river source to river mouth, do not necessarily take the shortest path. For alluvial streams, straight and braided rivers have very low sinuosity and flow directly down hill, while meandering rivers flow from side to side across a valley. Bedrock rivers typically flow in either a fractal pattern, or a pattern that is determined by weaknesses in the bedrock, such as faults, fractures, or more erodible layers.

Rate

Volumetric flow rate, also known as discharge, volume flow rate, and rate of water flow, is the volume of water which passes through a given cross-section of the river channel per unit time. It is typically measured in cubic metres per second (cumec) or cubic feet per second (cfs), where 1 m^3/s = 35.51 ft^3/s; it is sometimes also measured in litres or gallons per second.

Volumetric flow rate can be thought of as the mean velocity of the flow through a given cross-section, times that cross-sectional area. Mean velocity can be approximated through the use of the Law of the Wall. In general, velocity increases with the depth (or hydraulic radius) and slope of the river channel, while the cross-sectional area scales with the depth and the width: the double-counting of depth shows the importance of this variable in determining the discharge through the channel.

Fluvial Erosion

In the youthful stage, the river is subject to vertical erosion, deepening the valley. Hydraulic action loosens and dislodges the rock. The rivers load further erodes its banks and the river bed. Over time, this will deepen the river bed and create steeper sides which are then weathered.

The steepened nature of the banks causes the sides of the valley to move downslope causing the valley to become V-shaped.

Waterfalls also form in the youthful river valley.

Waterfalls usually form where a band of hard rock lies next to a layer of soft rock (easier to erode). Differential erosion occurs as the river can erode the soft rock easier than the hard rock, this leaves the hard rock more elevated and stands out from the river below. Hydraulic action and abrasion are what erodes the soft rock and the water to fall down to the river bed. A plunge pool forms at the bottom and deepens as a result of hydraulic action and abrasion.

Sediment Yield

Sediment yield is the total quantity of particulate matter (suspended or bedload) reaching the outlet of a drainage basin over a fixed time frame. Yield is usually expressed as kilograms per square kilometre per year. Sediment delivery processes are affected by a myriad of factors such as drainage area size, basin slope, climate, sediment type (lithology), vegetation cover, and human land use/management practices. The theoretical concept of the 'sediment delivery ratio' (ratio between yield and total amount of sediment eroded) captures the fact that not all of the sediment is eroded within a certain catchment that reaches out to the outlet (due to, for example, deposition on floodplains). Such storage opportunities are typically increased in catchments of larger size, thus leading to a lower yield and sediment delivery ratio.

Frozen river in Alaska.

Management

River bank repair.

Rivers are often managed or controlled to make them more useful or less disruptive to human activity.

- Dams or weirs may be built to control the flow, store water, or extract energy.
- Levees, known as dikes in Europe, may be built to prevent river water from flowing on floodplains or floodways.
- Canals connect rivers to one another for water transfer or navigation.
- River courses may be modified to improve navigation, or straightened to increase the flow rate.

River management is a continuous activity as rivers tend to 'undo' the modifications made by people. Dredged channels silt up, sluice mechanisms deteriorate with age, levees and dams may suffer seepage or catastrophic failure. The benefits sought through managing rivers may often be offset by the social and economic costs of mitigating the bad effects of such management. As an example, in parts of the developed world, rivers have been confined within channels to free up flat flood-plain land for development. Floods can inundate such development at high financial cost and often with loss of life.

Rivers are increasingly managed for habitat conservation, as they are critical for many aquatic and riparian plants, resident and migratory fishes, waterfowl, birds of prey, migrating birds, and many mammals.

WETLAND

Freshwater swamp forest in Bangladesh.

Peat bogs are freshwater wetlands that develop in areas with standing water and low soil fertility.

A wetland is a distinct ecosystem that is inundated by water, either permanently or seasonally, where oxygen-free processes prevail. The primary factor that distinguishes wetlands from other land forms or water bodies is the characteristic vegetation of aquatic plants, adapted to the unique hydric soil. Wetlands play a number of functions, including water purification, water storage, processing of carbon and other nutrients, stabilization of shorelines, and support of plants and animals. Wetlands are also considered the most biologically diverse of all ecosystems, serving as home to a wide range of plant and animal life. Whether any individual wetland performs these functions, and the degree to which it performs them, depends on characteristics of that wetland and the lands and waters near it. Methods for rapidly assessing these functions, wetland ecological health, and general wetland condition have been developed in many regions and have contributed to wetland conservation partly by raising public awareness of the functions and the ecosystem services some wetlands provide.

Marshes develop along the edges of rivers and lakes.

Wetlands occur naturally on every continent. The main wetland types are swamp, marsh, bog, and fen; sub-types include mangrove forest, carr, pocosin, floodplains, mire, vernal pool, sink, and many others. Many peatlands are wetlands. The water in wetlands is either freshwater, brackish, or saltwater. Wetlands can be tidal (inundated by tides) or non-tidal. The largest wetlands include the Amazon River basin, the West Siberian Plain, the Pantanal in South America, and the Sundarbans in the Ganges-Brahmaputra delta.

The UN Millennium Ecosystem Assessment determined that environmental degradation is more prominent within wetland systems than any other ecosystem on Earth.

Constructed wetlands are used to treat municipal and industrial wastewater as well as stormwater runoff. They may also play a role in water-sensitive urban design.

Technical Definitions

Sunrise at Viru Bog, Estonia.

A wetland is "an ecosystem that arises when inundation by water produces soils dominated by anaerobic and aerobic processes, which, in turn, forces the biota, particularly rooted plants, to adapt to flooding." There are four main kinds of wetlands – marsh, swamp, bog and fen (bogs and fens being types of mires). Some experts also recognize wet meadows and aquatic ecosystems as additional wetland types. The largest wetlands in the world include the swamp forests of the Amazon and the peatlands of Siberia.

Ramsar Convention Definition

Under the Ramsar international wetland conservation treaty, wetlands are defined as follows:

- Wetlands are areas of marsh, fen, peatland or water, whether natural or artificial, permanent or temporary, with water that is static or flowing, fresh, brackish or salt, including areas of marine water the depth of which at low tide does not exceed six metres.
- Wetlands may incorporate riparian and coastal zones adjacent to the wetlands, and islands or bodies of marine water deeper than six metres at low tide lying within the wetlands.

Regional Definitions

Although the general definition given above applies around the world, each county and region tends to have its own definition for legal purposes. In the United States, wetlands are defined as "those areas that are inundated or saturated by surface or groundwater at a frequency and duration sufficient to support, and that under normal circumstances do support, a prevalence of vegetation typically adapted for life in saturated soil conditions. Wetlands generally include swamps, marshes, bogs and similar areas". This definition has been used in the enforcement of the Clean Water Act. Some US states, such as Massachusetts and New York, have separate definitions that may differ from the federal government's.

- In the United States Code, the term wetland is defined "as land that (A) has a predominance of hydric soils, (B) is inundated or saturated by surface or groundwater at a frequency and duration sufficient to support a prevalence of hydrophytic vegetation typically adapted for life in saturated soil conditions and (C) under normal circumstances supports a prevalence of such vegetation." Related to this legal definitions, the term "normal circumstances" are conditions expected to occur during the wet portion of the growing season under normal climatic conditions (not unusually dry or unusually wet), and in the absence of significant disturbance. It is not uncommon for a wetland to be dry for long portions of the growing season. Wetlands can be dry during the dry season and abnormally dry periods during the wet season, but under normal environmental conditions the soils in a wetland will be saturated to the surface or inundated such that the soils become anaerobic, and those conditions will persist through the wet portion of the growing season.

Ecology

The most important factor producing wetlands is flooding. The duration of flooding or prolonged soil saturation by groundwater determines whether the resulting wetland has aquatic, marsh or swamp vegetation. Other important factors include fertility, natural disturbance, competition, herbivory, burial and salinity. When peat accumulates, bogs and fens arise.

Characteristics

Wetlands vary widely due to local and regional differences in topography, hydrology, vegetation, and other factors, including human involvement.

Hydrology

Wetland hydrology is associated with the spatial and temporal dispersion, flow, and physio-chemical attributes of surface and ground water in its reservoirs. Based on hydrology, wetlands can be categorized as riverine (associated with streams), lacustrine (associated with lakes and reservoirs), and palustrine (isolated). Sources of hydrological flows into wetlands are predominantly precipitation, surface water, and groundwater. Water flows out of wetlands by evapotranspiration, surface runoff, and subsurface water outflow. Hydrodynamics (the movement of water through and from a wetland) affects hydro-periods (temporal fluctuations in water levels) by controlling the water balance and water storage within a wetland.

Landscape characteristics control wetland hydrology and hydrochemistry. The O_2 and CO_2 concentrations of water depend on temperature and atmospheric pressure. Hydrochemistry within wetlands is determined by the pH, salinity, nutrients, conductivity, soil composition, hardness, and the sources of water. Water chemistry of wetlands varies across landscapes and climatic regions. Wetlands are generally minerotrophic with the exception of bogs.

Bogs receive most of their water from the atmosphere; therefore, their water usually has low mineral ionic composition. In contrast, groundwater has a higher concentration of dissolved nutrients and minerals.

The water chemistry of fens ranges from low pH and low minerals to alkaline with high accumulation of calcium and magnesium because they acquire their water from precipitation as well as ground water.

Role of Salinity

Salinity has a strong influence on wetland water chemistry, particularly in wetlands along the coast. and in regions with large precipitation deficits. In non-riverine wetlands, natural salinity is regulated by interactions between ground and surface water, which may be influenced by human activity.

Soil

Carbon is the major nutrient cycled within wetlands. Most nutrients, such as sulfur, phosphorus, carbon, and nitrogen are found within the soil of wetlands. Anaerobic and aerobic respiration in the soil influences the nutrient cycling of carbon, hydrogen, oxygen, and nitrogen, and the solubility of phosphorus thus contributing to the chemical variations in its water. Wetlands with low pH and saline conductivity may reflect the presence of acid sulfates and wetlands with average salinity levels can be heavily influenced by calcium or magnesium. Biogeochemical processes in wetlands are determined by soils with low redox potential. Wetland soils are identified by redoxymorphic mottles or low chroma, as determined by the Munsell Color System.

Biota

The biota of a wetland system includes its flora and fauna as described below. The most important factor affecting the biota is the duration of flooding. Other important factors include fertility and salinity. In fens, species are highly dependent on water chemistry. The chemistry of water flowing into wetlands depends on the source of water and the geological material in which it flows through as well as the nutrients discharged from organic matter in the soils and plants at higher elevations in slope wetlands. Biota may vary within a wetland due to season or recent flood regimes.

Flora

Bud of *Nelumbo nucifera*, an aquatic plant.

There are four main groups of hydrophytes that are found in wetland systems throughout the world.

Submerged wetland vegetation can grow in saline and fresh-water conditions. Some species have underwater flowers, while others have long stems to allow the flowers to reach the surface. Submerged species provide a food source for native fauna, habitat for invertebrates, and also possess filtration capabilities. Examples include seagrasses and eelgrass.

Floating water plants or floating vegetation is usually small, like arrow arum (*Peltandra virginica*).

Trees and shrubs, where they comprise much of the cover in saturated soils, qualify those areas in most cases as swamps. The upland boundary of swamps is determined partly by water levels. This can be affected by dams Some swamps can be dominated by a single species, such as silver maple swamps around the Great Lakes. Others, like those of the Amazon basin, have large numbers of different tree species. Examples include cypress (*Taxodium*) and mangrove.

Fauna

Fish are more dependent on wetland ecosystems than any other type of habitat. Seventy-five percent of the United States' commercial fish and shellfish stocks depend solely on estuaries to survive. Tropical fish species need mangroves for critical hatchery and nursery grounds and the coral reef system for food.

Many species of frogs live in wetlands, while others visit them each year to lay eggs.

Amphibians such as frogs need both terrestrial and aquatic habitats in which to reproduce and feed. While tadpoles control algal populations, adult frogs forage on insects. Frogs are used as an indicator of ecosystem health due to their thin skin which absorbs both nutrient and toxins from the surrounding environment resulting in an above average extinction rate in unfavorable and polluted environmental conditions.

Snapping turtles are one of the many kinds of turtles found in wetlands.

Reptiles such as alligators and crocodiles are common in wetlands of some regions.

Alligators occur in fresh water along with the fresh water species of the crocodile.The Florida Everglades is the only place in the world where both crocodiles and alligators coexist. The saltwater crocodile inhabits estuaries and mangroves and can be seen in the coastline bordering the Great Barrier Reef in Australia. Snakes, lizards and turtles also can be seen throughout wetlands. Snapping turtles are one of the many kinds of turtles found in wetlands. Birds, particularly waterfowl and wading birds, use wetlands extensively.

Mammals include numerous small and medium-sized species such as voles, bats, and platypus in addition to large herbivorous and apex species such as the beaver, coypu, swamp rabbit, Florida panther, and moose. Wetlands attract many mammals due to abundant seeds, berries, and other vegetation components, as well as abundant populations of prey such as invertebrates, small reptiles and amphibians.

Insects and invertebrates total more than half of the 100,000 known animal species in wetlands. Insects and invertebrates can be submerged in the water or soil, on the surface, and in the atmosphere.

Algae

Algae are diverse water plants that can vary in size, color, and shape. Algae occur naturally in habitats such as inland lakes, inter-tidal zones, and damp soil and provide a dedicated food source for many animals, including some invertebrates, fish, turtles, and frogs. There are three main groups of algae:

- Plankton are algae which are microscopic, free-floating algae. This algae is so tiny that on average, if 50 of these microscopic algae were lined up end-to-end, it would only measure one millimetre. Plankton are the basis of the food web and are responsible for primary production in the ocean using photosynthesis to make food.
- Filamentous algae are long strands of algae cells that form floating mats.
- Chara and Nitella algae are upright algae that look like a submerged plant with roots.

Climates

Temperature

Wetlands are located in every climatic zone. Temperatures vary greatly depending on the location of the wetland. Many of the world's wetlands are in temperate zones, midway between the North or South Pole and the equator. In these zones, summers are warm and winters are cold, but temperatures are not extreme. In a subtropical zone wetland, such as one along the Gulf of Mexico, a typical temperature might be 11 °C (52 °F). Wetlands in the tropics are much warmer for a larger portion of the year. Wetlands on the Arabian Peninsula can reach temperatures exceeding 50 °C (122 °F) and would therefore be subject to rapid evaporation. In northeastern Siberia, which has a polar climate, wetland temperatures can be as low as −50 °C (−58 °F). Peatlands insulate the permafrost in subarctic regions, thus delaying or preventing thawing of permafrost during summer, as well as inducing the formation of permafrost.

Wetlands contrast the hot, arid landscape around Middle Spring, Fish Springs National Wildlife Refuge, Utah.

Precipitation

The amount of precipitation a wetland receives varies widely according to its area. Wetlands in Wales, Scotland, and western Ireland typically receive about 1,500 mm (59 in) per year. In some places in Southeast Asia, where heavy rains occur, they can receive up to 10,000 mm (390 in). In some drier regions, wetlands exist where as little as 180 mm (7.1 in) precipitation occurs each year.

Temporal Variation

- Perennial systems.
- Seasonal systems.
- Episodic (periodic or intermittent) system of the down.
- Surface flow may occur in some segments, with subsurface flow in other segments.
- Ephemeral (short-lived) systems.
- Migratory species.

Uses of Wetlands

Depending partly on a wetland's geographic and topographic location, the functions it performs can support multiple ecosystem services, values, or benefits. United Nations Millennium Ecosystem Assessment and Ramsar Convention described wetlands as a whole to be of biosphere significance and societal importance in the following areas, for example:

- Water storage (flood control).

- Groundwater replenishment.
- Shoreline stabilisation and storm protection.
- Water purification.
- Reservoirs of biodiversity.
- Pollination.
- Wetland products.
- Cultural values.
- Recreation and tourism.
- Climate change mitigation and adaptation.

According to the Ramsar Convention:

> The economic worth of the ecosystem services provided to society by intact, naturally functioning wetlands is frequently much greater than the perceived benefits of converting them to 'more valuable' intensive land use – particularly as the profits from unsustainable use often go to relatively few individuals or corporations, rather than being shared by society as a whole.

Unless otherwise cited, ecosystem services information is based on the following series of references.

To replace these wetland ecosystem services, enormous amounts of money would need to be spent on water purification plants, dams, levees, and other hard infrastructure, and many of the services are impossible to replace.

Water Storage (Flood Control)

Major wetland type: floodplain and closed-depression wetlands.

Storage reservoirs and flood protection: The wetland system of floodplains is formed from major rivers downstream from their headwaters. "The floodplains of major rivers act as natural storage reservoirs, enabling excess water to spread out over a wide area, which reduces its depth and speed. Wetlands close to the headwaters of streams and rivers can slow down rainwater runoff and spring snowmelt so that it doesn't run straight off the land into water courses. This can help prevent sudden, damaging floods downstream." Notable river systems that produce large spans of floodplain include the Nile River, the Niger river inland delta, [the Zambezi River flood plain], [the Okavango River inland delta], [the Kafue River flood plain][the Lake Bangweulu flood plain] (Africa), Mississippi River (USA), Amazon River (South America), Yangtze River (China), Danube River (Central Europe) and Murray-Darling River (Australia).

Human impact: Converting wetlands to upland through drainage and development forces adjoining or downstream water channels into narrower corridors. This accelerates watershed hydrologic response to storm events and this increases the need in some cases for alternative means of flood control. That is because the newly formed channels must manage the same amount of precipitation, causing flood peaks to be [higher or deeper] and floodwaters to travel faster.

Water management engineering developments in the past century have degraded these wetlands through the construction of artificial embankments. These constructions may be classified as dykes, bunds, levees, weirs, barrages and dams but serve the single purpose of concentrating water into a select source or area. Wetland water sources that were once spread slowly over a large, shallow area are pooled into deep, concentrated locations. Loss of wetland floodplains results in more severe and damaging flooding. Catastrophic human impact in the Mississippi River floodplains was seen in death of several hundred individuals during a levee breach in New Orleans caused by Hurricane Katrina. Ecological catastrophic events from human-made embankments have been noticed along the Yangtze River floodplains since the middle of the river has become prone to more frequent and damaging flooding. Some of these events include the loss of riparian vegetation, a 30% loss of the vegetation cover throughout the river's basin, a doubling of the percentage of the land affected by soil erosion, and a reduction in reservoir capacity through siltation build-up in floodplain lakes.

Groundwater Replenishment

Major wetland type: marsh, swamp, and subterranean karst and cave hydrological systems.

The surface water which is the water visibly seen in wetland systems only represents a portion of the overall water cycle which also includes atmospheric water and groundwater. Wetland systems are directly linked to groundwater and a crucial regulator of both the quantity and quality of water found below the ground. Wetland systems that are made of permeable sediments like limestone or occur in areas with highly variable and fluctuating water tables especially have a role in groundwater replenishment or water recharge. Sediments that are porous allow water to filter down through the soil and overlying rock into aquifers which are the source of 95% of the world's drinking water. Wetlands can also act as recharge areas when the surrounding water table is low and as a discharge zone when it is too high. Karst (cave) systems are a unique example of this system and are a connection of underground rivers influenced by rain and other forms of precipitation. These wetland systems are capable of regulating changes in the water table on upwards of 130 m (430 ft).

Human impact: Groundwater is an important source of water for drinking and irrigation of crops. Over 1 billion people in Asia and 65% of the public water sources in Europe source 100% of their water from groundwater. Irrigation is a massive use of groundwater with 80% of the world's groundwater used for agricultural production.

Unsustainable abstraction of groundwater has become a major concern. In the Commonwealth of Australia, water licensing is being implemented to control use of water in major agricultural regions. On a global scale, groundwater deficits and water scarcity is one of the most pressing concerns facing the 21st century.

Shoreline Stabilization and Storm Protection

Tidal and inter-tidal wetland systems protect and stabilize coastal zones. Coral reefs provide a protective barrier to coastal shoreline. Mangroves stabilize the coastal zone from the interior and will migrate with the shoreline to remain adjacent to the boundary of the water. The main conservation benefit these systems have against storms and storm surges is the ability to reduce the speed and height of waves and floodwaters.

Human impact: The sheer number of people who live and work near the coast is expected to grow immensely over the next fifty years. From an estimated 200 million people that currently live in low-lying coastal regions, the development of urban coastal centers is projected to increase the population by fivefold within 50 years. The United Kingdom has begun the concept of managed coastal realignment. This management technique provides shoreline protection through restoration of natural wetlands rather than through applied engineering. In East Asia, reclamation of coastal wetlands has resulted in widespread transformation of the coastal zone, and up to 65% of coastal wetlands have been destroyed by coastal development. One analysis using the impact of hurricanes versus storm protection provided naturally by wetlands projected the value of this service at US$33,000/hectare/year.

Water Purification

Wetland types: floodplain, closed-depression wetlands, mudflat, salt marsh, mangroves.

Nutrient retention: Wetlands cycle both sediments and nutrients balancing terrestrial and aquatic ecosystems. A natural function of wetland vegetation is the up-take, storage, and (for nitrate) the removal of nutrients found in runoff from the surrounding soil and water. In many wetlands, nutrients are retained until plants die or are harvested by animals or humans and taken to another location, or until microbial processes convert soluble nutrients to a gas as is the case with nitrate.

Sediment and heavy metal traps: Precipitation and surface runoff induces soil erosion, transporting sediment in suspension into and through waterways. These sediments move towards larger and more sizable waterways through a natural process that moves water towards oceans. All types of sediments which may be composed of clay, sand, silt, and rock can be carried into wetland systems through this process. Wetland vegetation acts as a physical barrier to slow water flow and trap sediment for short or long periods of time. Suspended sediment often contains heavy metals that are retained when wetlands trap the sediment. In some cases, certain metals are taken up through wetland plant stems, roots, and leaves. Many floating plant species, for example, can absorb and filter heavy metals. Water hyacinth (*Eichhornia crassipes*), duckweed (*Lemna*) and water fern (*Azolla*) store iron and copper commonly found in wastewater. Many fast-growing plants rooted in the soils of wetlands such as cattail (*Typha*) and reed (*Phragmites*) also aid in the role of heavy metal up-take. Animals such as the oyster can filter more than 200 litres (53 US gal) of water per day while grazing for food, removing nutrients, suspended sediments, and chemical contaminants in the process. On the other hand, some types of wetlands facilitate the mobilization and bioavailability of mercury (another heavy metal), which in its methyl mercury form increases the risk of bioaccumulation in fish important to animal food webs and harvested for human consumption.

Capacity: The ability of wetland systems to store or remove nutrients and trap sediment and associated metals is highly efficient and effective but each system has a threshold. An overabundance of nutrient input from fertilizer run-off, sewage effluent, or non-point pollution will cause eutrophication. Upstream erosion from deforestation can overwhelm wetlands making them shrink in size and cause dramatic biodiversity loss through excessive sedimentation load. Retaining high levels of metals in sediments is problematic if the sediments become resuspended or oxygen and pH levels change at a future time. The capacity of wetland vegetation to store heavy metals depends on the particular metal, oxygen and pH status of wetland sediments and overlying water, water flow rate (detention time), wetland size, season, climate, type of plant, and other factors.

Human impact: The capacity of a wetland to store sediment, nutrients, and metals can be diminished if sediments are compacted such as by vehicles or heavy equipment, or are regularly tilled. Unnatural changes in water levels and water sources also can affect the water purification function. If water purification functions are impaired, excessive loads of nutrients enter waterways and cause eutrophication. This is of particular concern in temperate coastal systems. The main sources of coastal eutrophication are industrially made nitrogen, which is used as fertilizer in agricultural practices, as well as septic waste runoff. Nitrogen is the limiting nutrient for photosynthetic processes in saline systems, however in excess, it can lead to an overproduction of organic matter that then leads to hypoxic and anoxic zones within the water column.[Without oxygen, other organisms cannot survive, including economically important finfish and shellfish species.

Examples: An example of how a natural wetland is used to provide some degree of sewage treatment is the East Kolkata Wetlands in Kolkata, India. The wetlands cover 125 square kilometres (48 sq mi), and are used to treat Kolkata's sewage. The nutrients contained in the wastewater sustain fish farms and agriculture.

Constructed Wetlands

The function of most natural wetland systems is not to manage wastewater. However, their high potential for the filtering and the treatment of pollutants has been recognized by environmental engineers that specialize in the area of wastewater treatment. These constructed wetland systems are highly controlled environments that intend to mimic the occurrences of soil, flora, and microorganisms in natural wetlands to aid in treating wastewater effluent. Constructed wetlands can be used to treat raw sewage, storm water, agricultural and industrial effluent. They are constructed with flow regimes, micro-biotic composition, and suitable plants in order to produce the most efficient treatment process. Other advantages of constructed wetlands are the control of retention times and hydraulic channels. The most important factors of constructed wetlands are the water flow processes combined with plant growth.

Constructed wetland in an ecological settlement in Flintenbreite near Lübeck, Germany.

Constructed wetland systems can be surface flow systems with only free-floating macrophytes, floating-leaved macrophytes, or submerged macrophytes; however, typical free water surface systems are usually constructed with emergent macrophytes. Subsurface flow-constructed wetlands with a vertical or a horizontal flow regime are also common and can be integrated into urban areas as they require relatively little space.

Reservoirs of Biodiversity

Wetland systems' rich biodiversity is becoming a focal point at International Treaty Conventions and within the World Wildlife Fund organization due to the high number of species present in wetlands, the small global geographic area of wetlands, the number of species which are endemic to wetlands, and the high productivity of wetland systems. Hundred of thousands of animal species, 20,000 of them vertebrates, are living in wetland systems. The discovery rate of fresh water fish is at 200 new species per year. The impact of maintaining biodiversity is seen at the local level through job creation, sustainability, and community productivity. A good example is the Lower Mekong basin which runs through Cambodia, Laos, and Vietnam. Supporting over 55 million people, the sustainability of the region is enhanced through wildlife tours. The U.S. state of Florida has estimated that US$1.6 billion was generated in state revenue from recreational activities associated with wildlife.

Biodiverse river basins: The Amazon holds 3,000 species of freshwater fish species within the boundaries of its basin, whose function it is to disperse the seeds of trees. One of its key species, the Piramutaba catfish, *Brachyplatystoma vaillantii*, migrates more than 3,300 km (2,100 mi) from its nursery grounds near the mouth of the Amazon River to its spawning grounds in Andean tributaries, 400 m (1,300 ft) above sea level, distributing plants seed along the route.

Productive intertidal zones: Intertidal mudflats have a level of productivity similar to that of some wetlands even while possessing a low number of species. The abundance of invertebrates found within the mud are a food source for migratory waterfowl.

Critical life-stage habitat: Mudflats, saltmarshes, mangroves, and seagrass beds have high levels of both species richness and productivity, and are home to important nursery areas for many commercial fish stocks.

Genetic diversity: Populations of many species are confined geographically to only one or a few wetland systems, often due to the long period of time that the wetlands have been physically isolated from other aquatic sources. For example, the number of endemic species in Lake Baikal in Russia classifies it as a hotspot for biodiversity and one of the most biodiverse wetlands in the entire world. Evidence from a research study by Mazepova *et al.* suggest that the number of crustacean species endemic to Baikal Lake (over 690 species and subspecies) exceeds the number of the same groups of animals inhabiting all the fresh water bodies of Eurasia together. Its 150 species of free-living Platyhelminthes alone is analogous to the entire number in all of Eastern Siberia. The 34 species and subspecies number of Baikal sculpins is more than twice the number of the analogous fauna that inhabits Eurasia. One of the most exciting discoveries was made by A. V. Shoshin who registered about 300 species of free-living nematodes using only six near-shore sampling localities in the Southern Baikal. "If we will take into consideration, that about 60% of the animals can be found nowhere else except Baikal, it may be assumed that the lake may be the biodiversity center of the Eurasian continent."

Human impact: Biodiversity loss occurs in wetland systems through land use changes, habitat destruction, pollution, exploitation of resources, and invasive species. Vulnerable, threatened, and endangered species number at 17% of waterfowl, 38% of fresh-water dependent mammals, 33% of freshwater fish, 26% of freshwater amphibians, 72% of freshwater turtles, 86% of marine turtles, 43% of crocodilians and 27% of coral reef-building species. Introduced hydrophytes in different wetland systems can have devastating results. The introduction of water hyacinth, a native plant of South America into Lake Victoria in East Africa as well as duckweed into non-native areas of Queensland, Australia, have overtaken entire wetland systems suffocating the wetlands and reducing the diversity of other plants and animals. This is largely due to their phenomenal growth rate and ability to float and grow on the surface of the water.

Wetland Products and Productivity

Wetland productivity is linked to the climate, wetland type, and nutrient availability. Low water and occasional drying of the wetland bottom during droughts (dry marsh phase) stimulate plant recruitment from a diverse seed bank and increase productivity by mobilizing nutrients. In contrast, high water during deluges (lake marsh phase) causes turnover in plant populations and creates greater interspersion of element cover and open water, but lowers overall productivity. During a cover cycle that ranges from open water to complete vegetation cover, annual net primary productivity may vary 20-fold. The grasses of fertile floodplains such as the Nile produce the highest yield including plants such as *Arundo donax* (giant reed), *Cyperus papyrus* (papyrus), *Phragmites* (reed) and *Typha* (cattail, bulrush).

Wetlands naturally produce an array of vegetation and other ecological products that can be harvested for personal and commercial use. The most significant of these is fish which have all or part of their life-cycle occur within a wetland system. Fresh and saltwater fish are the main source of protein for one billion people and comprise 15% of an additional two billion people's diets. In addition, fish generate a fishing industry that provides 80% of the income and employment to residents in developing countries. Another food staple found in wetland systems is rice, a popular grain that is consumed at the rate of one fifth of the total global calorie count. In Bangladesh, Cambodia and Vietnam, where rice paddies are predominant on the landscape, rice consumption reach 70%. Some native wetland plants in the Caribbean and Australia are harvested sustainably for medicinal compounds; these include the red mangrove (*Rhizophora mangle*) which possesses antibacterial, wound-healing, anti-ulcer effects, and antioxidant properties.

Food converted to sweeteners and carbohydrates include the sago palm of Asia and Africa (cooking oil), the nipa palm of Asia (sugar, vinegar, alcohol, and fodder) and honey collection from mangroves. More than supplemental dietary intake, this produce sustains entire villages. Coastal Thailand villages earn the key portion of their income from sugar production while the country of Cuba relocates more than 30,000 hives each year to track the seasonal flowering of the mangrove *Avicennia*.

Other mangrove-derived products:

- Fuelwood.
- Salt (produced by evaporating seawater).
- Animal fodder.
- Traditional medicines (e.g. from mangrove bark).
- Fibers for textiles.
- Dyes and tannins.

Human impact: Over-fishing is the major problem for sustainable use of wetlands. Concerns are developing over certain aspects of farm fishing, which uses natural waterways to harvest fish for human consumption and pharmaceuticals. This practice has become especially popular in Asia and the South Pacific. Its impact upon much larger waterways downstream has negatively affected many small island developing states.

Aquaculture is continuing to develop rapidly throughout the Asia-Pacific region specifically in China with world holdings in Asia equal to 90% of the total number of aquaculture farms and 80% of its global value. Some aquaculture has eliminated massive areas of wetland through practices seen such as in the shrimp farming industry's destruction of mangroves. Even though the damaging impact of large scale shrimp farming on the coastal ecosystem in many Asian countries has been widely recognized for quite some time now, it has proved difficult to check in absence of other employment avenues for people engaged in such occupation. Also burgeoning demand for shrimps globally has provided a large and ready market for the produce.

Threats to rice fields mainly stem from inappropriate water management, introduction of invasive alien species, agricultural fertilizers, pesticides, and land use changes. Industrial-scale production of palm oil threatens the biodiversity of wetland ecosystems in parts of southeast Asia, Africa, and other developing countries.

Over-exploitation of wetland products can occur at the community level as is sometimes seen throughout coastal villages of Southern Thailand where each resident may obtain for themselves every consumable of the mangrove forest (fuelwood, timber, honey, resins, crab, and shellfish) which then becomes threatened through increasing population and continual harvest.

Additional Functions and uses of Wetlands

Some types of wetlands can serve as fire breaks that help slow the spread of minor wildfires. Larger wetland systems can influence local precipitation patterns. Some boreal wetland systems in catchment headwaters may help extend the period of flow and maintain water temperature in connected downstream waters. Pollination services are supported by many wetlands which may provide the only suitable habitat for pollinating insects, birds, and mammals in highly developed areas. It is likely that wetlands have other functions whose benefits to society and other ecosystems have yet to be discovered.

Wetlands and Climate Change

Wetlands perform two important functions in relation to climate change. They have mitigation effects through their ability to sink carbon, converting a greenhouse gas (carbon dioxide) to solid plant material through the process of photosynthesis, and also through their ability to store and regulate water. Wetlands store approximately 44.6 million tonnes of carbon per year globally. In salt marshes and mangrove swamps in particular, the average carbon sequestration rate is 210 g CO_2 m^{-2} y^{-1} while peatlands sequester approximately.

20–30 g CO_2 m^{-2} y^{-1}. Coastal wetlands, such as tropical mangroves and some temperate salt marshes, are known to be sinks for carbon that otherwise contributes to climate change in its gaseous forms (carbon dioxide and methane). The ability of many tidal wetlands to store carbon and minimize methane flux from tidal sediments has led to sponsorship of blue carbon initiatives that are intended to enhance those processes.

However, depending on their characteristics, some wetlands are a significant source of methane emissions and some are also emitters of nitrous oxide which is a greenhouse gas with a global warming potential 300 times that of carbon dioxide and is the dominant ozone-depleting substance emitted in the 21st century. Excess nutrients mainly from anthropogenic sources have been shown to significantly increase the N_2O fluxes from wetland soils through denitrification and nitrification processes. A study in the intertidal region of a New England salt marsh showed that excess levels of nutrients might increase N_2O emissions rather than sequester them.

Nitrous Oxide Fluxes from Different Wetland Soils

Wetland type	Location	N_2O flux (μmol N_2O m^{-2} h^{-1})
Mangrove	Shenzhen and Hong Kong	0.14 – 23.83
Mangrove	Muthupet, South India	0.41 – 0.77
Mangrove	Bhitarkanika, East India	0.20 – 4.73
Mangrove	Pichavaram, South India	0.89 – 1.89
Mangrove	Queensland, Australia	−0.045 – 0.32
Mangrove	South East Queensland, Australia	0.091 – 1.48
Mangrove	Southwest coast, Puerto Rico	0.12 – 7.8
Mangrove	Isla Magueyes, Puerto Rico	0.05 – 1.4
Salt marsh	Chesapeake Bay, US	0.005 – 0.12
Salt marsh	Maryland, US	0.1
Salt marsh	North East China	0.1 – 0.16
Salt marsh	Biebrza, Poland	−0.07 – 0.06
Salt marsh	Netherlands	0.82 – 1.64
Salt marsh	Baltic Sea	−0.13
Salt marsh	Massachusetts, US	−2.14 – 1.27

Data on nitrous oxide fluxes from wetlands in the southern hemisphere are lacking, as are ecosystem-based studies including the role of dominant organisms that alter sediment biogeochemistry. Aquatic invertebrates produce ecologically-relevant nitrous oxide emissions due to ingestion of denitrifying bacteria that live within the subtidal sediment and water column and thus may also be influencing nitrous oxide production within some wetlands.

Peatswamps in Southeast Asia

In Southeast Asia, peatswamp forests and soils are being drained, burnt, mined, and overgrazed, contributing severely to climate change. As a result of peat drainage, the organic carbon that was built up over thousands of years and is normally under water is suddenly exposed to the air. It decomposes and turns into carbon dioxide (CO_2), which is released into the atmosphere. Peat fires cause the same process to occur and in addition create enormous clouds of smoke that cross international borders, such as happens every year in Southeast Asia. While peatlands constitute only 3% of the world's land area, their degradation produces 7% of all fossil fuel CO_2 emissions.

Through the building of dams, Wetlands International is halting the drainage of peatlands in Southeast Asia, hoping to mitigate CO_2 emissions. Concurrent wetland restoration techniques include reforestation with native tree species as well as the formation of community fire brigades. This sustainable approach can be seen in central Kalimantan and Sumatra, Indonesia.

Wetland Disturbance

Wetlands, the functions and services they provide as well as their flora and fauna, can be affected by several types of disturbances. The disturbances (sometimes termed stressors or alterations) can be human-associated or natural, direct or indirect, reversible or not, and isolated or cumulative. When exceeding levels or patterns normally found within wetlands of a particular class in a particular region, the predominant ones include the following:

- Enrichment/Eutrophication.
- Organic Loading and Reduced Dissolved Oxygen.
- Contaminant Toxicity.
- Acidification.
- Salinization.
- Sedimentation.
- Altered Solar Input (Turbidity/Shade).
- Vegetation Removal.
- Thermal Alteration.

- Dehydration/Aridification.
- Inundation/Flooding.
- Habitat Fragmentation.
- Other Human Presence.

Disturbances can be further categorized as follows:

Minor Disturbance

Stress that maintains ecosystem integrity.

Moderate Disturbance

Ecosystem integrity is damaged but can recover in time without assistance.

Impairment or Severe Disturbance

Human intervention may be needed in order for ecosystem to recover.

Just a few of the many sources of these disturbances are:

- Drainage.
- Development.
- Over-grazing.
- Mining.
- Unsustainable water use.

They can be manifested partly as:

- Water scarcity.
- Impacts to Endangered species.
- Disruption of wildlife breeding grounds.
- Imbalance in sediment load and nutrient filtration.

Water Chemistry

Anthropogenic nitrogen inputs to aquatic systems have drastically effected the dissolved nitrogen content of wetlands, introducing higher nutrient availability which leads to eutrophication., Due to the low dissolved oxygen (DO) content, and relatively low nutrient balance of wetland environments, they are very susceptible to alterations in water chemistry. Key factors that are assessed to determine water quality include:

- Major Anion Analysis: (HCO_3^-,Cl^-,NO_3^-,SO_4^{2-}).
- Major Cation Analysis (Ca^{2+}, Mg^{2+}, Na^+, K^+).
- pH.
- Conductivity increases with more dissolved ions in the water.
- Turbidity.
- Dissolved Oxygen.
- Temperature.
- Total Dissolved Solids.

These chemical factors can be used to quantify wetland disturbances, and often provide information as to whether a wetland is surface water fed or groundwater fed due to the different ion characteristics of the two water sources. Wetlands are adept at impacting the water chemistry of streams or water bodies that interact with them, and can withdraw ions that result from water pollution such as acid mine drainage or urban runoff.

Conservation

Wetlands have historically been the victim of large draining efforts for real estate development, or flooding for use as recreational lakes or hydropower generation. Some of the world's most important agricultural areas are wetlands that have been converted to farmland. Since the 1970s, more focus has been put on preserving wetlands for their natural function yet by 1993 half the world's wetlands had been drained.

In order to maintain wetlands and sustain their functions, alterations and disturbances that are outside the normal range of variation should be minimized.

Fog rising over the Mukri bog near Mukri, Estonia. The bog has an area of 2,147 hectares (5,310 acres) and has been protected since 1992.

Balancing Wetland Conservation with the Needs of People

Wetlands are vital ecosystems that provide livelihoods for the millions of people who live in and around them. The Millennium Development Goals (MDGs) called for different sectors to join forces to secure wetland environments in the context of sustainable development and improving human wellbeing. A three-year project carried out by Wetlands International in partnership with the International Water Management Institute found that it is possible to conserve wetlands while improving the livelihoods of people living among them. Case studies conducted in Malawi and Zambia looked at how dambos – wet, grassy valleys or depressions where water seeps to the surface – can be farmed sustainably to improve livelihoods. Mismanaged or overused dambos often become degraded, however, using a knowledge exchange between local farmers and environmental managers, a protocol was developed using soil and water management practices. Project outcomes included a high yield of crops, development of sustainable farming techniques, and adequate water management generating enough water for use as irrigation. Before the project, there were cases where people had died from starvation due to food shortages. By the end of it, many more people had access to enough water to grow vegetables. A key achievement was that villagers had secure food supplies during long, dry months. They also benefited in other ways: nutrition was improved by growing a wider range of crops, and villagers could also invest in health and education by selling produce and saving money.

Ramsar Convention

The Convention on Wetlands of International Importance, especially as Waterfowl Habitat, or Ramsar Convention, is an international treaty designed to address global concerns regarding wetland loss and degradation. The primary purposes of the treaty are to list wetlands of international importance and to promote their wise use, with the ultimate goal of preserving the world's wetlands. Methods include restricting access to the majority portion of wetland areas, as well as educating the public to combat the misconception that wetlands are wastelands. The Convention works closely with five International Organisation Partners. These are: Birdlife International, the IUCN, the International Water Management Institute, Wetlands International and the World Wide Fund for Nature. The partners provide technical expertise, help conduct or facilitate field studies and provide financial support. The IOPs also participate regularly as observers in all meetings of the Conference of the Parties and the Standing Committee and as full members of the Scientific and Technical Review Panel.

Valuation

The value of a wetland to local communities, as well as the value of wetland systems generally to the earth and to humankind, is one of the most important valuations that can be conducted for sustainable development. This typically involves first mapping a region's wetlands, then assessing the functions and ecosystem services the wetlands provide individually and cumulatively, and evaluating that information to prioritize or rank individual wetlands or wetland types for conservation, management, restoration, or development. Over a longer period, it requires keeping inventories of known wetlands and monitoring a representative sample of the wetlands to determine changes due to both natural and human factors. Such a valuation process is used to educate decision-makers such as governments of the importance of particular wetlands within their jurisdiction.

Assessment

Rapid assessment methods are used to score, rank, rate, or categorize various functions, ecosystem services, species, communities, levels of disturbance, and/or ecological health of a wetland or group of wetlands. This is often done to prioritize particular wetlands for conservation (avoidance) or to determine the degree to which loss or alteration of wetland functions should be compensated, such as by restoring degraded wetlands elsewhere or providing additional protections to existing wetlands. Rapid assessment methods are also applied before and after a wetland has been restored or altered, to help monitor or predict the effects of those actions on various wetland functions and the services they provide. Assessments are typically considered to be "rapid" when they require only a single visit to the wetland lasting less than one day, which in some cases may include interpretation of aerial imagery and GIS analyses of existing spatial data, but not detailed post-visit laboratory analyses of water or biological samples. Due to time and cost constraints, the levels of various wetland functions or other attributes are usually not measured directly but rather are estimated relative to other assessed wetlands in a region, using observation-based variables, sometimes called "indicators", that are hypothesized or known to predict performance of the specified functions or attributes.

To achieve consistency among persons doing the assessment, rapid methods present indicator variables as questions or checklists on standardized data forms, and most methods standardize the scoring or rating procedure that is used to combine question responses into estimates of the levels of specified functions relative to the levels estimated in other wetlands ("calibration sites") assessed previously in a region. Rapid assessment methods, partly because they often use dozens of indicators pertaining to conditions surrounding a wetland as well as within the wetland itself, aim to provide estimates of wetland functions and services that are more accurate and repeatable than simply describing a wetland's class type. A need for wetland assessments to be rapid arises mostly when government agencies set deadlines for decisions affecting a wetland, or when the number of wetlands needing information on their functions or condition is large.

In North America and a few other countries, standardized rapid assessment methods for wetlands have a long history, having been developed, calibrated, tested, and applied to varying degrees in several different regions and wetland types since the 1970s. However, few rapid assessment methods have been fully validated. Done correctly, validation is a very expensive endeavor that involves comparing rankings of a series of wetlands based on results from rapid assessment methods with rankings based on less rapid and considerably more costly, multi-visit, detailed measurements of levels of the same functions or other attributes in the same series of wetlands.

Monitoring

A wetland needs to be monitored over time to assess whether it is functioning at an ecologically sustainable level or whether it is becoming degraded. Degraded wetlands will suffer a loss in water quality, loss of sensitive species, and aberrant functioning of soil geochemical processes.

Mapping

Practically, many natural wetlands are difficult to monitor from the ground as they quite often are difficult to access and may require exposure to dangerous plants and animals as well as diseases borne by insects or other invertebrates. Therefore, mapping using aerial imagery is one effective tool to monitor a wetland, especially a large wetland, and can also be used to monitor the status of numerous wetlands throughout a watershed or region. Many remote sensing methods can be used to map wetlands. Remote-sensing technology permits the acquisition of timely digital data on a repetitive basis. This repeat coverage allows wetlands, as well as the adjacent land-cover and land-use types, to be monitored seasonally and/or annually. Using digital data provides a standardized data-collection procedure and an opportunity for data integration within a geographic information system. Traditionally, Landsat 5 Thematic Mapper (TM), Landsat 7 Enhanced Thematic Mapper Plus (ETM+), and the SPOT 4 and 5 satellite systems have been used for this purpose. More recently, however, multispectral IKONOS and QuickBird data, with spatial resolutions of 4 by 4 m (13 by 13 ft) and 2.44 by 2.44 m (8.0 by 8.0 ft), respectively, have been shown to be excellent sources of data when mapping and monitoring smaller wetland habitats and vegetation communities.

For example, Detroit Lakes Wetland Management District assessed area wetlands in Michigan, USA, using remote sensing. Through using this technology, satellite images were taken over a large geographic area and extended period. In addition, using this technique was less costly and time-consuming compared to the older method using visual interpretation of aerial photographs. In comparison, most aerial photographs also require experienced interpreters to extract information based on structure and texture while the interpretation of remote sensing data only requires analysis of one characteristic (spectral).

However, there are a number of limitations associated with this type of image acquisition. Analysis of wetlands has proved difficult because to obtain the data it is often linked to other purposes such as the analysis of land cover or land use.

Further Improvements

Methods to develop a classification system for specific biota of interest could assist with technological advances that will allow for identification at a very high accuracy rate. The issue of the cost and expertise involved in remote sensing technology is still a factor hindering further advancements in image acquisition and data processing. Future improvements in current wetland vegetation mapping could include the use of more recent and better geospatial data when it is available.

Restoration

Restoration and restoration ecologists intend to return wetlands to their natural trajectory by aiding directly with the natural processes of the ecosystem. These direct methods vary with respect to the degree of physical manipulation of the natural environment and each are associated with different levels of restoration. Restoration is needed after disturbance or perturbation of a wetland. Disturbances include exogenous factors such as flooding or drought. Other external damage may be anthropogenic disturbance caused by clear-cut harvesting of trees, oil and gas extraction, poorly defined infrastructure installation, over grazing of livestock, ill-considered recreational activities, alteration of wetlands including dredging, draining, and filling, and other negative human impacts. Disturbance puts different levels of stress on an environment depending on the type and duration of disturbance. There is no one way to restore a wetland and the level of restoration required will be based on the level of disturbance although, each method of restoration does require preparation and administration.

Levels of Restoration

Factors Influencing Selected Approach may Include

- Budget.
- Time scale limitations.
- Project goals.
- Level of disturbance.
- Landscape and ecological constraints.
- Political and administrative agendas.
- Socioeconomic priorities.

Prescribed Natural Regeneration

There are no biophysical manipulation and the ecosystem is left to recover based on the process of succession alone. The focus of this method is to eliminate and prevent further disturbance from occurring. In order for this type of restoration to be effective and successful there must be prior research done to understand the probability that the wetland will recover with this method. Otherwise, some biophysical manipulation may be required to enhance the rate of succession to an acceptable level determined by the project managers and ecologists. This is likely to be the first method of approach for the lowest level of disturbance being that it is the least intrusive and least costly.

- Assisted Natural Regeneration.

    There are some biophysical manipulations however they are non-intrusive. Example methods that are not limited to wetlands include prescribed burns to small areas, promotion of site specific soil microbiota and plant growth using nucleation planting whereby plants radiate from an initial planting site, and promotion of niche diversity or increasing the range of niches to promote use by a variety of different species. These methods can make it easier for the natural species to flourish by removing competition from their environment and can speed up the process of succession.

- Partial Reconstruction.

    Here there is a mix between natural regeneration and manipulated environmental control. These manipulations may require some engineering and more invasive biophysical manipulation including ripping of subsoil, agrichemical applications such as herbicides and insecticides, laying of mulch, mechanical seed dispersal, and tree planting on a large scale. In these circumstances the wetland is impaired and without human assistance it would not recover within an acceptable period of time determined by ecologists. Again these methods of restoration will have to be considered on a site by site basis as each site will require a different approach based on levels of disturbance and ecosystem dynamics.

Complete Reconstruction

The most expensive and intrusive method of reconstruction requiring engineering and ground up reconstruction. Because there is a redesign of the entire ecosystem it is important that the natural trajectory of the ecosystem be considered and that the plant species will eventually return the ecosystem towards its natural trajectory.

Important Considerations

- Constructed wetlands can take 10–100 years to fully resemble the vegetative composition of a natural wetland.
- Artificial wetlands do not have hydric soil. The soil has very low levels of organic carbon and total nitrogen compared to natural wetland systems, and this reduces the performance of several functions.

- Organic matter added to degraded natural wetlands can in some cases help restore their productivity.

Legislation

International Efforts

- Ramsar convention.
- North American waterfowl management plan.

Canadian National Efforts

- The federal policy on wetland conservation.
- Other individual provincial and territorial based policies.

List of Wetland Types

The following list is that used within Australia to classify wetland by type:

Marine and Coastal Zone Wetlands

- Marine waters—permanent shallow waters less than six metres deep at low tide; includes sea bays, straits.
- Subtidal aquatic beds; includes kelp beds, seagrasses, tropical marine meadows.
- Coral reefs.
- Rocky marine shores; includes rocky offshore islands, sea cliffs.
- Sand, shingle or pebble beaches; includes sand bars, spits, sandy islets.
- Intertidal mud, sand or salt flats.
- Intertidal marshes; includes saltmarshes, salt meadows, saltings, raised salt marshes, tidal brackish and freshwater marshes.
- Intertidal forested wetlands; includes mangrove swamps, nipa swamps, tidal freshwater swamp forests.
- Brackish to saline lagoons and marshes with one or more relatively narrow connections with the sea.
- Freshwater lagoons and marshes in the coastal zone.
- Non-tidal freshwater forested wetlands.

Inland Wetlands

- Permanent rivers and streams; includes waterfalls.
- Seasonal and irregular rivers and streams.
- Inland deltas (permanent).
- Riverine floodplains; includes river flats, flooded river basins, seasonally flooded grassland, savanna and palm savanna.
- Permanent freshwater lakes (> 8 ha); includes large oxbow lakes.
- Seasonal/intermittent freshwater lakes (> 8 ha), floodplain lakes.
- Permanent saline/brackish lakes.
- Seasonal/intermittent saline lakes.
- Permanent freshwater ponds (< 8 ha), marshes and swamps on inorganic soils; with emergent vegetation waterlogged for at least most of the growing season.
- Seasonal/intermittent freshwater ponds and marshes on inorganic soils; includes sloughs, potholes; seasonally flooded meadows, sedge marshes.
- Permanent saline/brackish marshes.
- Seasonal saline marshes.
- Shrub swamps; shrub-dominated freshwater marsh, shrub carr, alder thicket on inorganic soils.
- Freshwater swamp forest; seasonally flooded forest, wooded swamps; on inorganic soils.
- Peatlands; forest, shrub or open bogs.
- Alpine and tundra wetlands; includes alpine meadows, tundra pools, temporary waters from snow melt.
- Freshwater springs, oases and rock pools.
- Geothermal wetlands.
- Inland, subterranean karst wetlands.

Human-made Wetlands

- Water storage areas; reservoirs, barrages, hydro-electric dams, impoundments (generally > 8 ha).
- Ponds, including farm ponds, stock ponds, small tanks (generally < 8 ha).
- Aquaculture ponds; fish ponds, shrimp ponds.
- Salt exploitation; salt pans, salines.

- Excavations; gravel pits, borrow pits, mining pools.
- Wastewater treatment; sewage farms, settling ponds, oxidation basins.
- Irrigated land and irrigation channels; rice fields, canals, ditches.
- Seasonally flooded arable land, farm land.

OCEAN

Surface view of the Atlantic Ocean.

An ocean is a body of water that composes much of a planet's hydrosphere. On Earth, an ocean is one of the major conventional divisions of the World Ocean. These are, in descending order by area, the Pacific, Atlantic, Indian, Southern (Antarctic), and Arctic Oceans. The word "ocean" is often used interchangeably with "sea" in American English. Strictly speaking, a *sea* is a body of water (generally a division of the world ocean) partly or fully enclosed by land, though "the sea" refers also to the oceans.

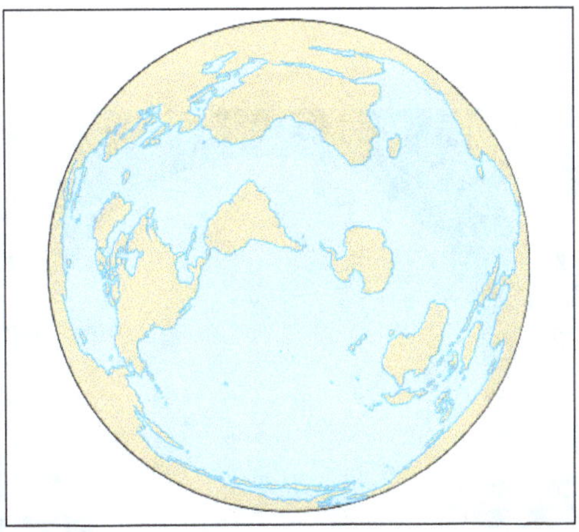

Saline water covers approximately 361,000,000 km2 (139,000,000 sq mi) and is customarily divided into several principal oceans and smaller seas, with the ocean covering approximately 71% of Earth's surface and 90% of the Earth's biosphere. The ocean contains 97% of Earth's water, and oceanographers have stated that less than 5% of the World Ocean has been explored. The total volume is approximately 1.35 billion cubic kilometers (320 million cu mi) with an average depth of nearly 3,700 meters (12,100 ft).

As the world ocean is the principal component of Earth's hydrosphere, it is integral to life, forms part of the carbon cycle, and influences climate and weather patterns. The World Ocean is the habitat of 230,000 known species, but because much of it is unexplored, the number of species that exist in the ocean is much larger, possibly over two million. The origin of Earth's oceans is unknown; oceans are thought to have formed in the Hadean eon and may have been the impetus for the emergence of life.

Extraterrestrial oceans may be composed of water or other elements and compounds. The only confirmed large stable bodies of extraterrestrial surface liquids are the lakes of Titan, although there is evidence for the existence of oceans elsewhere in the Solar System. Early in their geologic histories, Mars and Venus are theorized to have had large water oceans. The Mars ocean hypothesis suggests that nearly a third of the surface of Mars was once covered by water, and a runaway greenhouse effect may have boiled away the global ocean of Venus. Compounds such as salts and ammonia dissolved in water lower its freezing point so that water might exist in large quantities in extraterrestrial environments as brine or convecting ice. Unconfirmed oceans are speculated beneath the surface of many dwarf planets and natural satellites; notably, the ocean of the moon Europa is estimated to have over twice the water volume of Earth. The Solar System's giant planets are also thought to have liquid atmospheric layers of yet to be confirmed compositions. Oceans may also exist on exoplanets and exomoons, including surface oceans of liquid water within a circumstellar habitable zone. Ocean planets are a hypothetical type of planet with a surface completely covered with liquid.

Earth's Global Ocean

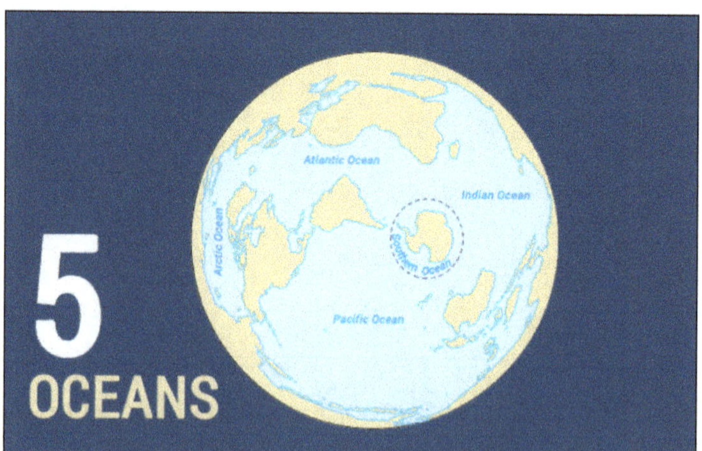

Various ways to divide the World Ocean.

Oceanic Divisions

Though generally described as several separate oceans, the global, interconnected body of salt water is sometimes referred to as the World Ocean or global ocean. The concept of a continuous body of water with relatively free interchange among its parts is of fundamental importance to oceanography.

1. Epipelagic zone: surface – 200 meters deep 2. Mesopelagic zone: 200 m – 1000 m 3. Bathypelagic zone: 1000 m – 4000 m 4. Abyssopelagic zone: 4000 m – 6000 m 5. Hadal zone (the trenches): 6000 m to the bottom of the ocean.

The major oceanic divisions – listed below in descending order of area and volume – are defined in part by the continents, various archipelagos, and other criteria.

	Ocean	Location	Area (km²) (%)	Volume (km³) (%)	Avg. depth (m)	Coastline (km)
1	Pacific Ocean	Separates Asia and Oceania from the Americas	168,723,000 *46.6*	669,880,000 *50.1*	3,970	135,663
2	Atlantic Ocean	Separates the Americas from Europe and Africa	85,133,000 *23.5*	310,410,900 *23.3*	3,646	111,866
3	Indian Ocean	Washes upon southern Asia and separates Africa and Australia	70,560,000 *19.5*	264,000,000 *19.8*	3,741	66,526
4	Southern Ocean	Sometimes considered an extension of the Pacific, Atlantic and Indian Oceans, which encircles Antarctica	21,960,000 *6.1*	71,800,000 *5.4*	3,270	17,968
5	Arctic Ocean	Sometimes considered a sea or estuary of the Atlantic, which covers much of the Arctic and washes upon northern North America and Eurasia	15,558,000 *4.3*	18,750,000 *1.4*	1,205	45,389

Seas and Bays

Sea	Location	Area (sq. km)	#
Arabian Sea	Between the Arabian peninsula and the Indian subcontinent	3,862,000	1
Bay of Bengal	Between the Indian subcontinent and the malaysia peninsula	2,173,000	2

NB: Volume, area, and average depth figures include NOAA ETOPO1 figures for marginal South China Sea.

Oceans are fringed by smaller, adjoining bodies of water such as seas, gulfs, bays, bights, and straits.

Global System

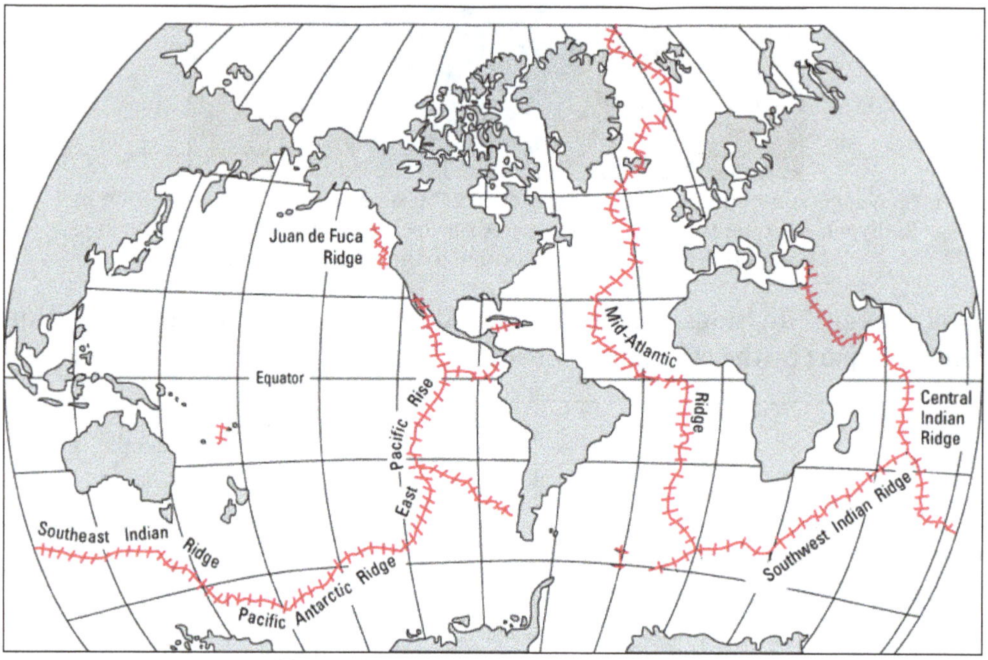

World Distribution of Mid-Oceanic Ridges; USGS.

The mid-ocean ridges of the world are connected and form a single global mid-oceanic ridge system that is part of every ocean and the longest mountain range in the world. The continuous mountain range is 65,000 km (40,000 mi) long (several times longer than the Andes, the longest continental mountain range).

Physical Properties

The total mass of the hydrosphere is about 1.4 quintillion tonnes (1.4×10^{18} long tons or 1.5×10^{18} short tons), which is about 0.023% of Earth's total mass. Less than 3% is freshwater; the rest is saltwater, almost all of which is in the ocean. The area of the

World Ocean is about 361.9 million square kilometers (139.7 million square miles), which covers about 70.9% of Earth's surface, and its volume is approximately 1.335 billion cubic kilometers (320.3 million cubic miles). This can be thought of as a cube of water with an edge length of 1,101 kilometers (684 mi). Its average depth is about 3,688 meters (12,100 ft), and its maximum depth is 10,994 meters (6.831 mi) at the Mariana Trench. Nearly half of the world's marine waters are over 3,000 meters (9,800 ft) deep. The vast expanses of deep ocean (anything below 200 meters or 660 feet) cover about 66% of Earth's surface. This does not include seas not connected to the World Ocean, such as the Caspian Sea.

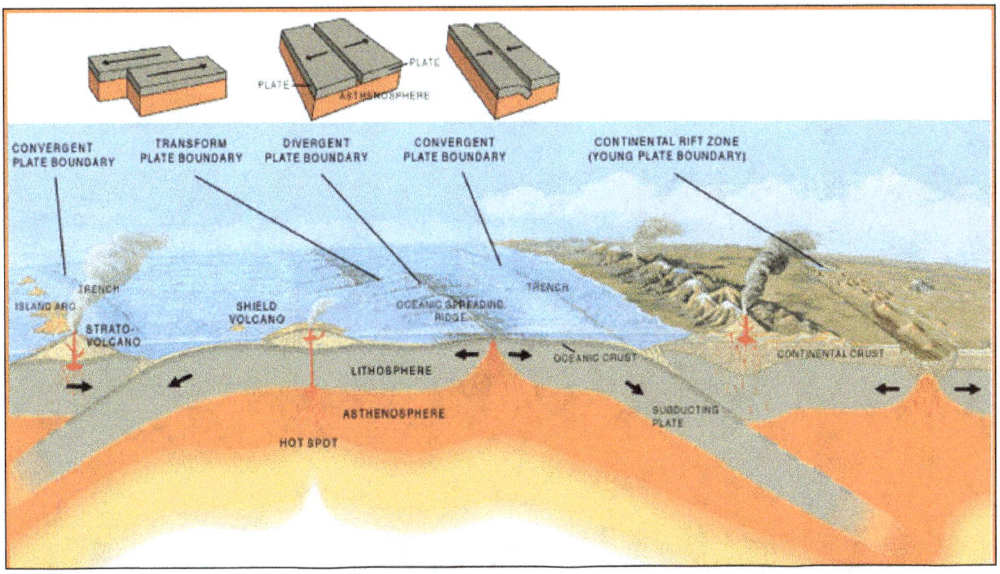

Three main types of plate boundaries.

The bluish ocean color is a composite of several contributing agents. Prominent contributors include dissolved organic matter and chlorophyll. Mariners and other seafarers have reported that the ocean often emits a visible glow which extends for miles at night. In 2005, scientists announced that for the first time, they had obtained photographic evidence of this glow. It is most likely caused by bioluminescence.

Oceanic Zones

Oceanographers divide the ocean into different vertical zones defined by physical and biological conditions. The pelagic zone includes all open ocean regions, and can be divided into further regions categorized by depth and light abundance. The photic zone includes the oceans from the surface to a depth of 200 m; it is the region where photosynthesis can occur and is, therefore, the most biodiverse. Because plants require photosynthesis, life found deeper than the photic zone must either rely on material sinking from above or find another energy source. Hydrothermal vents are the primary source of energy in what is known as the aphotic zone (depths exceeding 200 m). The pelagic part of the photic zone is known as the epipelagic.

The pelagic part of the aphotic zone can be further divided into vertical regions according to temperature. The mesopelagic is the uppermost region. Its lowermost boundary is at a thermocline of 12 °C (54 °F), which, in the tropics generally lies at 700–1,000 meters (2,300–3,300 ft). Next is the bathypelagic lying between 10 and 4 °C (50 and 39 °F), typically between 700–1,000 meters (2,300–3,300 ft) and 2,000–4,000 meters (6,600–13,100 ft), lying along the top of the abyssal plain is the abyssopelagic, whose lower boundary lies at about 6,000 meters (20,000 ft). The last zone includes the deep oceanic trench, and is known as the hadalpelagic. This lies between 6,000–11,000 meters (20,000–36,000 ft) and is the deepest oceanic zone.

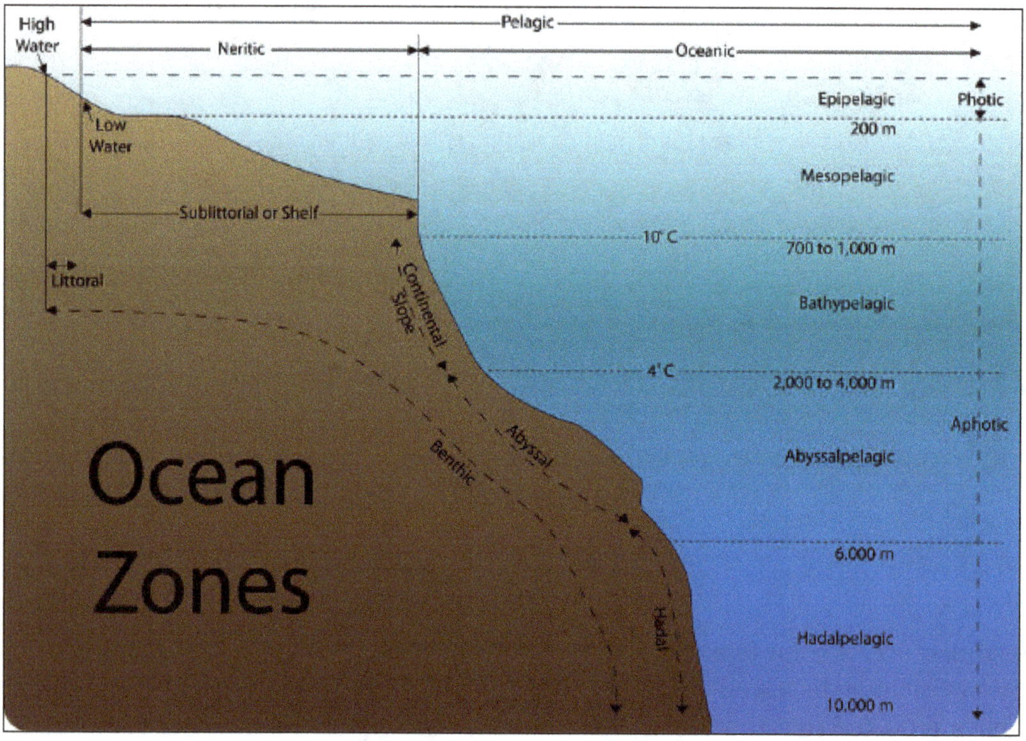

The major oceanic zones, based on depth and biophysical conditions.

The benthic zones are aphotic and correspond to the three deepest zones of the deep-sea. The bathyal zone covers the continental slope down to about 4,000 meters (13,000 ft). The abyssal zone covers the abyssal plains between 4,000 and 6,000 m. Lastly, the hadal zone corresponds to the hadalpelagic zone, which is found in oceanic trenches.

The pelagic zone can be further subdivided into two subregions: the neritic zone and the oceanic zone. The neritic zone encompasses the water mass directly above the continental shelves whereas the oceanic zone includes all the completely open water.

In contrast, the littoral zone covers the region between low and high tide and represents the transitional area between marine and terrestrial conditions. It is also known as the intertidal zone because it is the area where tide level affects the conditions of the region.

If a zone undergoes dramatic changes in temperature with depth, it contains a thermocline. The tropical thermocline is typically deeper than the thermocline at higher latitudes. Polar waters, which receive relatively little solar energy, are not stratified by temperature and generally lack a thermocline because surface water at polar latitudes are nearly as cold as water at greater depths. Below the thermocline, water is very cold, ranging from −1 °C to 3 °C. Because this deep and cold layer contains the bulk of ocean water, the average temperature of the world ocean is 3.9 °C. If a zone undergoes dramatic changes in salinity with depth, it contains a halocline. If a zone undergoes a strong, vertical chemistry gradient with depth, it contains a chemocline.

The halocline often coincides with the thermocline, and the combination produces a pronounced pycnocline.

Exploration

The deepest point in the ocean is the Mariana Trench, located in the Pacific Ocean near the Northern Mariana Islands. Its maximum depth has been estimated to be 10,971 meters (35,994 ft). The British naval vessel *Challenger II* surveyed the trench in 1951 and named the deepest part of the trench the "Challenger Deep". In 1960, the Trieste successfully reached the bottom of the trench, manned by a crew of two men.

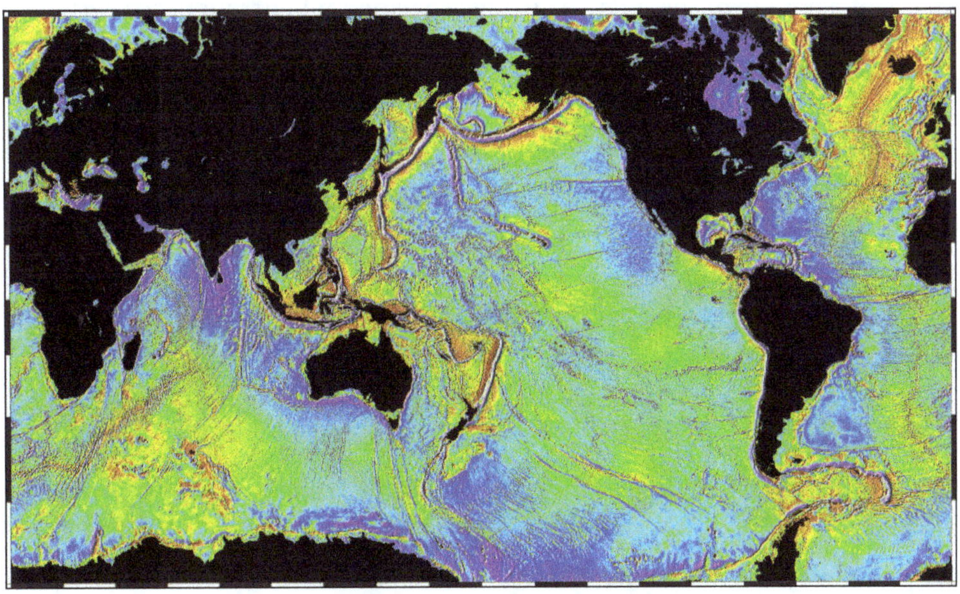

Map of large underwater features.

Oceanic Maritime Currents

Oceanic maritime currents have different origins. Tidal currents are in phase with the tide, hence are quasiperiodic; they may form various knots in certain places, most notably around headlands. Non-periodic currents have for origin the waves, wind and different densities.

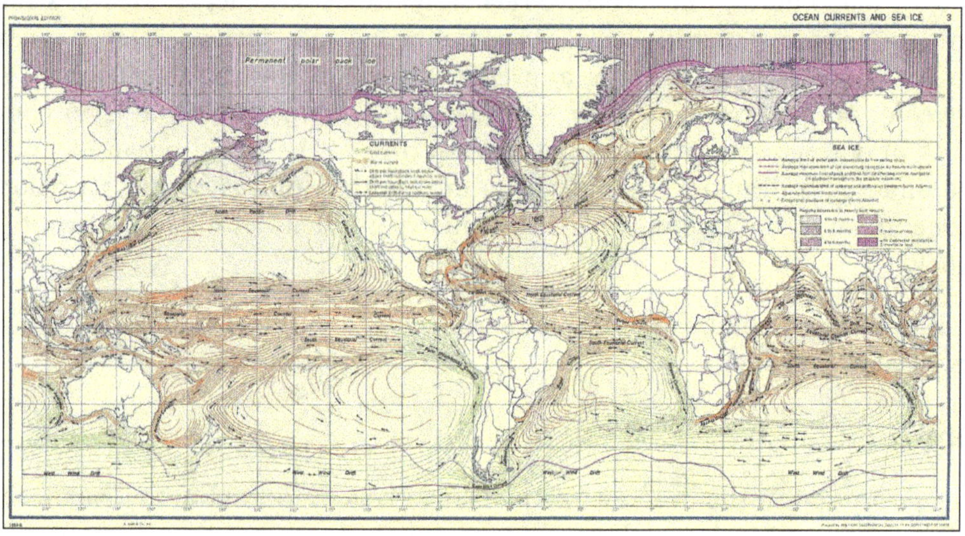

Oceanic surface currents.

The wind and waves create surface currents (designated as "drift currents"). These currents can decompose in one quasi-permanent current (which varies within the hourly scale) and one movement of Stokes drift under the effect of rapid waves movement (at the echelon of a couple of seconds). The quasi-permanent current is accelerated by the breaking of waves, and in a lesser governing effect, by the friction of the wind on the surface.

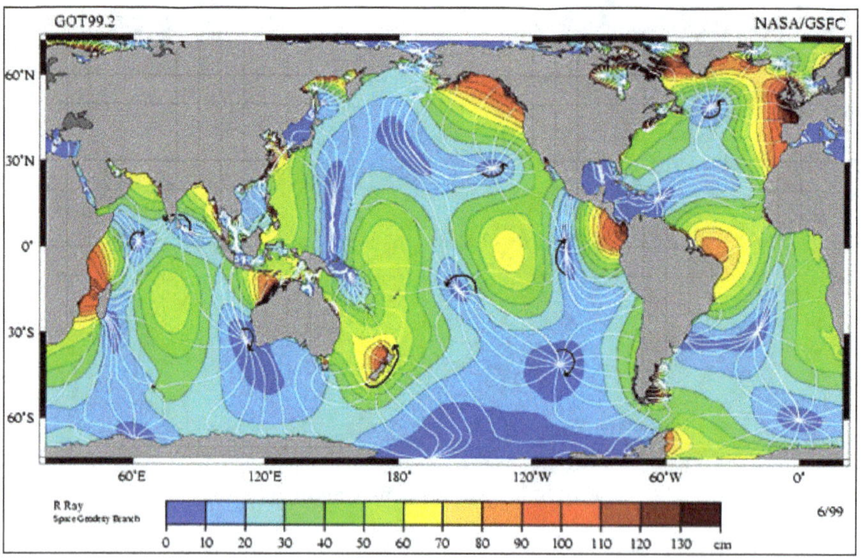

Amphidromic points showing the direction of tides per incrementation periods along with resonating directions of wavelength movements.

This acceleration of the current takes place in the direction of waves and dominant wind.

Accordingly, when the sea depth increases, the rotation of the earth changes the direction of currents in proportion with the increase of depth, while friction lowers their speed. At a certain sea depth, the current changes direction and is seen inverted in the opposite direction with current speed becoming null: known as the Ekman spiral. The influence of these currents is mainly experienced at the mixed layer of the ocean surface, often from 400 to 800 meters of maximum depth. These currents can considerably alter, change and are dependent on the various yearly seasons. If the mixed layer is less thick (10 to 20 meters), the quasi-permanent current at the surface adopts an extreme oblique direction in relation to the direction of the wind, becoming virtually homogeneous, until the Thermocline.

In the deep however, maritime currents are caused by the temperature gradients and the salinity between water density masses.

In littoral zones, breaking waves are so intense and the depth measurement so low, that maritime currents reach often 1 to 2 knots.

Climate

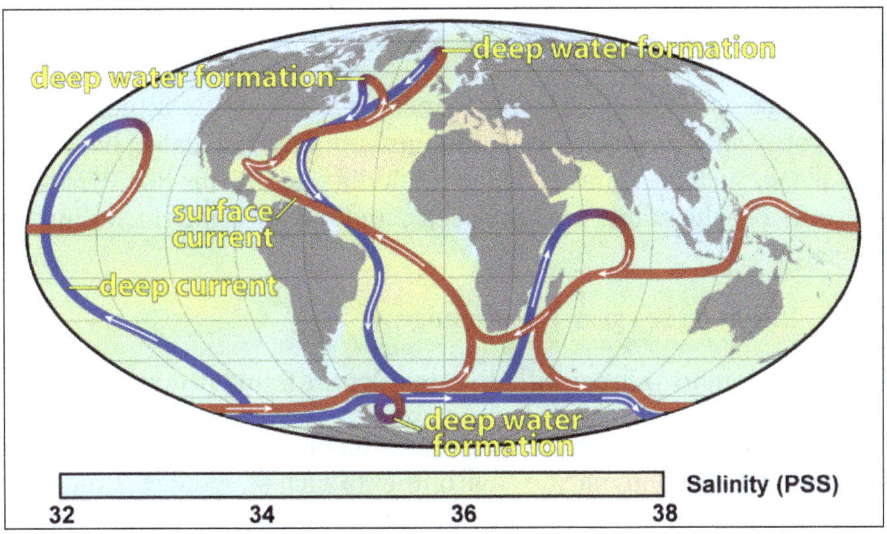

A map of the global thermohaline circulation; blue represent deep-water currents, whereas red represent surface currents.

Ocean currents greatly affect Earth's climate by transferring heat from the tropics to the polar regions. Transferring warm or cold air and precipitation to coastal regions, winds may carry them inland. Surface heat and freshwater fluxes create global density gradients that drive the thermohaline circulation part of large-scale ocean circulation. It plays an important role in supplying heat to the polar regions, and thus in sea ice regulation. Changes in the thermohaline circulation are thought to have significant impacts on Earth's energy budget. In so far as the thermohaline circulation governs the rate at which deep waters reach the surface, it may also significantly influence atmospheric carbon dioxide concentrations.

The Antarctic Circumpolar Current encircles that continent, influencing the area's climate and connecting currents in several oceans.

One of the most dramatic forms of weather occurs over the oceans: tropical cyclones (also called "typhoons" and "hurricanes" depending upon where the system forms).

Biology

The ocean has a significant effect on the biosphere. Oceanic evaporation, as a phase of the water cycle, is the source of most rainfall, and ocean temperatures determine climate and wind patterns that affect life on land. Life within the ocean evolved 3 billion years prior to life on land. Both the depth and the distance from shore strongly influence the biodiversity of the plants and animals present in each region.

As it "is" thought that life evolved in the ocean, the diversity of life is immense, including:

- Bacteria: Ubiquitous single-celled prokaryotes found throughout the world.
- Archaea: Prokaryotes distinct from bacteria, that inhabit many environments of the ocean, as well as many extreme environments.
- Algae: Algae is a "catch-all" term to include many photosynthetic, single-celled eukaryotes, such as green algae, diatoms, and dinoflagellates, but also multicellular algae, such as some red algae (including organisms like Pyropia, which is the source of the edible nori seaweed), and brown algae (including organisms like kelp).
- Plants: Including sea grasses, or mangroves.
- Fungi: Many marine fungi with diverse roles are found in oceanic environments.
- Animals: Most animal phyla have species that inhabit the ocean, including many that are only found in marine environments such as sponges, Cnidaria (such as corals and jellyfish), comb jellies, Brachiopods, and Echinoderms (such as sea urchins and sea stars). Many other familiar animal groups primarily live in the ocean, including cephalopods (includes octopus and squid), crustaceans (includes lobsters, crabs, and shrimp), fish, sharks, cetaceans (includes whales, dolphins, and porpoises).

In addition, many land animals have adapted to living a major part of their life on the oceans. For instance, seabirds are a diverse group of birds that have adapted to a life mainly on the oceans. They feed on marine animals and spend most of their lifetime on water, many only going on land for breeding. Other birds that have adapted to oceans as their living space are penguins, seagulls and pelicans. Seven species of turtles, the sea turtles, also spend most of their time in the oceans.

Gases

Table: Characteristics of oceanic gases.

Gas	Concentration of seawater, by mass (in parts per million), for the whole ocean	% Dissolved gas, by volume, in seawater at the ocean surface
Carbon dioxide (CO_2)	64 to 107	15%
Nitrogen (N_2)	10 to 18	48%
Oxygen (O_2)	0 to 13	36%

Table: Solubility of oceanic gases (in mL/L) with temperature at salinity of 33‰ and atmospheric pressure.

Temperature	O_2	CO_2	N_2
0 °C	8.14	8,700	14.47
10 °C	6.42	8,030	11.59
20 °C	5.26	7,350	9.65
30 °C	4.41	6,600	8.26

Surface

Table: Generalized characteristics of ocean surface by latitude.

Characteristic	Oceanic waters in polar regions	Oceanic waters in temperate regions	Oceanic waters in tropical regions
Precipitation vs. evaporation	P > E	P > E	E > P
Sea surface temperature in winter	−2 °C	5 to 20 °C	20 to 25 °C
Average salinity	28‰ to 32‰	35‰	35‰ to 37‰
Annual variation of air temperature	≤ 40ªC	10 °C	< 5 °C
Annual variation of water temperature	< 5ªC	10 °C	< 5 °C

Mixing Time

Table: Mean oceanic residence time for various constituents.

Constituent	Residence time (in years)
Iron (Fe)	200
Aluminum (Al)	600
Manganese (Mn)	1,300
Water (H_2O)	4,100
Silicon (Si)	20,000
Carbonate (CO_3^{2-})	110,000
Calcium (Ca^{2+})	1,000,000
Sulfate (SO_4^{2-})	11,000,000

Potassium (K^+)	12,000,000
Magnesium (Mg^{2+})	13,000,000
Sodium (Na^+)	68,000,000
Chloride (Cl^-)	100,000,000

Salinity

A zone of rapid salinity increase with depth is called a halocline. The temperature of maximum density of seawater decreases as its salt content increases. Freezing temperature of water decreases with salinity, and boiling temperature of water increases with salinity. Typical seawater freezes at around −2 °C at atmospheric pressure. If precipitation exceeds evaporation, as is the case in polar and temperate regions, salinity will be lower. If evaporation exceeds precipitation, as is the case in tropical regions, salinity will be higher. Thus, oceanic waters in polar regions have lower salinity content than oceanic waters in temperate and tropical regions.

Salinity can be calculated using the chlorinity, which is a measure of the total mass of halogen ions (includes fluorine, chlorine, bromine, and iodine) in seawater. By international agreement, the following formula is used to determine salinity:

$$\text{Salinity (in ‰)} = 1.80655 \times \text{Chlorinity (in ‰)}$$

The average chlorinity is about 19.2‰, and, thus, the average salinity is around 34.7‰

Absorption of Light

Table: Absorption of light in different wavelengths by ocean.

Color: Wavelength (nm)	Depth at which 99 percent of the wavelength is absorbed (in meters)	Percent absorbed in 1 meter of water
Ultraviolet (UV): 310	31	14.0
Violet (V): 400	107	4.2
Blue (B): 475	254	1.8
Green (G): 525	113	4.0
Yellow (Y): 575	51	8.7
Orange (O): 600	25	16.7
Red (R): 725	4	71.0
Infrared (IR): 800	3	82.0

Economic Value

Many of the world's goods are moved by ship between the world's seaports. Oceans are also the major supply source for the fishing industry. Some of the major harvests are shrimp, fish, crabs, and lobster.

Waves and Swell

The motions of the ocean surface, known as undulations or *waves*, are the partial and alternate rising and falling of the ocean surface. The series of mechanical waves that propagate along the interface between water and air is called swell.

Extraterrestrial Oceans

Although Earth is the only known planet with large stable bodies of liquid water on its surface and the only one in the Solar System, other celestial bodies are thought to have large oceans.

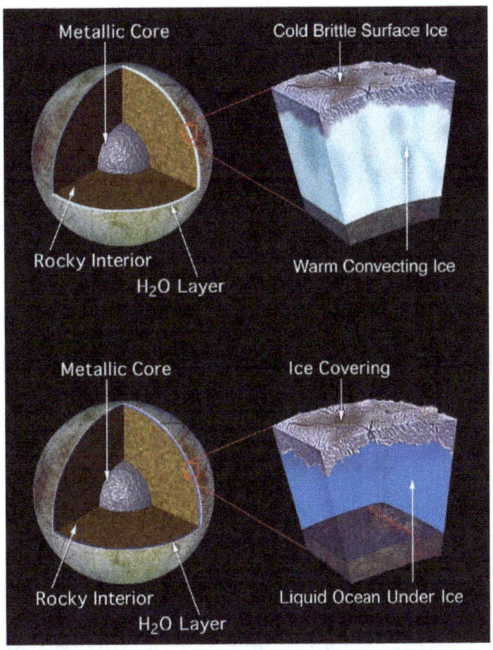

Two models for the composition of Europa predict a large subsurface ocean of liquid water. Similar models have been proposed for other celestial bodies in the Solar System.

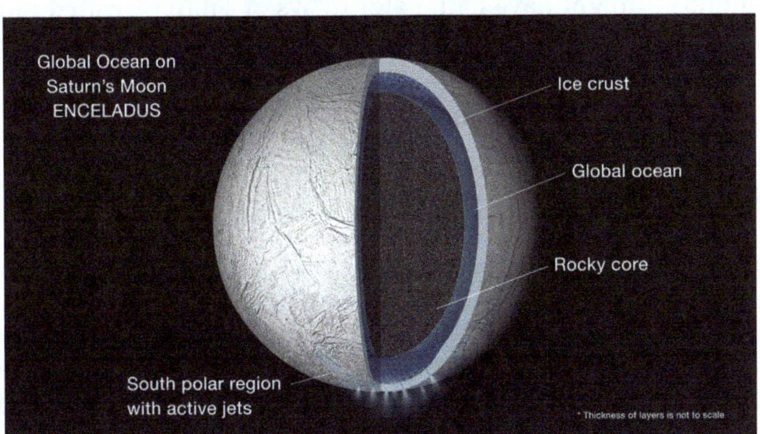

Artist's conception of subsurface ocean of Enceladus.

Planets

The gas giants, Jupiter and Saturn, are thought to lack surfaces and instead have a stratum of liquid hydrogen; however their planetary geology is not well understood. The possibility of the ice giants Uranus and Neptune having hot, highly compressed, supercritical water under their thick atmospheres has been hypothesised. Although their composition is still not fully understood, a 2006 study by Wiktorowicz and Ingersall ruled out the possibility of such a water "ocean" existing on Neptune, though some studies have suggested that exotic oceans of liquid diamond are possible.

The Mars ocean hypothesis suggests that nearly a third of the surface of Mars was once covered by water, though the water on Mars is no longer oceanic (much of it residing in the ice caps). The possibility continues to be studied along with reasons for their apparent disappearance. Astronomers now think that Venus may have had liquid water and perhaps oceans for over 2 billion years.

Natural Satellites

A global layer of liquid water thick enough to decouple the crust from the mantle is thought to be present on the natural satellites Titan, Europa, Enceladus and, with less certainty, Callisto, Ganymede and Triton. A magma ocean is thought to be present on Io. Geysers have been found on Saturn's moon Enceladus, possibly originating from an ocean about 10 kilometers (6.2 mi) beneath the surface ice shell. Other icy moons may also have internal oceans, or may once have had internal oceans that have now frozen.

Large bodies of liquid hydrocarbons are thought to be present on the surface of Titan, although they are not large enough to be considered oceans and are sometimes referred to as *lakes* or seas. The Cassini–Huygens space mission initially discovered only what appeared to be dry lakebeds and empty river channels, suggesting that Titan had lost what surface liquids it might have had. Later flybys of Titan provided radar and infrared images that showed a series of hydrocarbon lakes in the colder polar regions. Titan is thought to have a subsurface liquid-water ocean under the ice in addition to the hydrocarbon mix that forms atop its outer crust.

Dwarf Planets and Trans-Neptunian Objects

Ceres appears to be differentiated into a rocky core and icy mantle and may harbour a liquid-water ocean under its surface.

Not enough is known of the larger trans-Neptunian objects to determine whether they are differentiated bodies capable of supporting oceans, although models of radioactive decay suggest that Pluto, Eris, Sedna, and Orcus have oceans beneath solid icy crusts approximately 100 to 180 km thick.

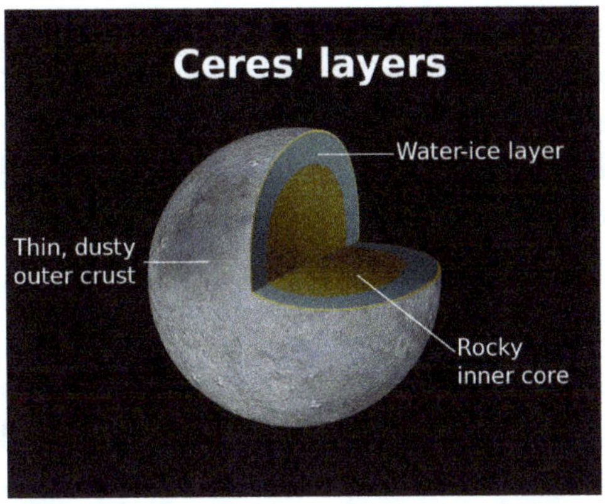

Diagram showing a possible internal structure of Ceres.

Extrasolar

Rendering of a hypothetical large extrasolar moon with surface liquid-water oceans.

Some planets and natural satellites outside the Solar System are likely to have oceans, including possible water ocean planets similar to Earth in the habitable zone or "liquid-water belt". The detection of oceans, even through the spectroscopy method, however is likely extremely difficult and inconclusive.

Theoretical models have been used to predict with high probability that GJ 1214 b, detected by transit, is composed of exotic form of ice VII, making up 75% of its mass, making it an ocean planet.

Other possible candidates are merely speculated based on their mass and position in the habitable zone include planet though little is actually known of their composition. Some scientists speculate Kepler-22b may be an "ocean-like" planet. Models have been proposed for Gliese 581 d that could include surface oceans. Gliese 436 b is speculated to have an ocean of "hot ice". Exomoons orbiting planets, particularly gas giants within their parent star's habitable zone may theoretically have surface oceans.

Terrestrial planets will acquire water during their accretion, some of which will be buried in the magma ocean but most of it will go into a steam atmosphere, and when the atmosphere cools it will collapse on to the surface forming an ocean. There will also be outgassing of water from the mantle as the magma solidifies—this will happen even for planets with a low percentage of their mass composed of water, so "super-Earth exoplanets may be expected to commonly produce water oceans within tens to hundreds of millions of years of their last major accretionary impact."

Non-water Surface Liquids

Oceans, seas, lakes and other bodies of liquids can be composed of liquids other than water, for example the hydrocarbon lakes on Titan. The possibility of seas of nitrogen on Triton was also considered but ruled out. There is evidence that the icy surfaces of the moons Ganymede, Callisto, Europa, Titan and Enceladus are shells floating on oceans of very dense liquid water or water–ammonia. Earth is often called *the* ocean planet because it is 70% covered in water. Extrasolar terrestrial planets that are extremely close to their parent star will be tidally locked and so one half of the planet will be a magma ocean. It is also possible that terrestrial planets had magma oceans at some point during their formation as a result of giant impacts. Hot Neptunes close to their star could lose their atmospheres via hydrodynamic escape, leaving behind their cores with various liquids on the surface. Where there are suitable temperatures and pressures, volatile chemicals that might exist as liquids in abundant quantities on planets include ammonia, argon, carbon disulfide, ethane, hydrazine, hydrogen, hydrogen cyanide, hydrogen sulfide, methane, neon, nitrogen, nitric oxide, phosphine, silane, sulfuric acid, and water.

Supercritical fluids, although not liquids, do share various properties with liquids. Underneath the thick atmospheres of the planets Uranus and Neptune, it is expected that these planets are composed of oceans of hot high-density fluid mixtures of water, ammonia and other volatiles. The gaseous outer layers of Jupiter and Saturn transition smoothly into oceans of supercritical hydrogen. The atmosphere of Venus is 96.5% carbon dioxide, which is a supercritical fluid at its surface.

GROUND WATER

Groundwater is the water present beneath Earth's surface in soil pore spaces and in the fractures of rock formations. A unit of rock or an unconsolidated deposit is called an aquifer when it can yield a usable quantity of water. The depth at which soil pore spaces or fractures and voids in rock become completely saturated with water is called the water table. Groundwater is recharged from and eventually flows to the surface naturally; natural discharge often occurs at springs and seeps, and can form oases or wetlands. Groundwater is also often withdrawn for agricultural, municipal, and industrial use by constructing and operating extraction wells. The study of the distribution and movement of groundwater is hydrogeology, also called groundwater hydrology.

Typically, groundwater is thought of as water flowing through shallow aquifers, but, in the technical sense, it can also contain soil moisture, permafrost (frozen soil), immobile water in very low permeability bedrock, and deep geothermal or oil formation water. Groundwater is hypothesized to provide lubrication that can possibly influence the movement of faults. It is likely that much of Earth's subsurface contains some water, which may be mixed with other fluids in some instances. Groundwater may not be confined only to Earth. The formation of some of the landforms observed on Mars may have been influenced by groundwater. There is also evidence that liquid water may also exist in the subsurface of Jupiter's moon Europa.

The entire surface water flow of the Alapaha River near Jennings, Florida going into a sinkhole leading to the Floridan Aquifer groundwater.

Groundwater is often cheaper, more convenient and less vulnerable to pollution than surface water. Therefore, it is commonly used for public water supplies. For example, groundwater provides the largest source of usable water storage in the United States, and California annually withdraws the largest amount of groundwater of all the states. Underground reservoirs contain far more water than the capacity of all surface reservoirs and lakes in the US, including the Great Lakes. Many municipal water supplies are derived solely from groundwater.

Polluted groundwater is less visible and more difficult to clean up than pollution in rivers and lakes. Groundwater pollution most often results from improper disposal of wastes on land. Major sources include industrial and household chemicals and garbage landfills, excessive fertilizers and pesticides used in agriculture, industrial waste lagoons, tailings and process wastewater from mines, industrial fracking, oil field brine pits, leaking underground oil storage tanks and pipelines, sewage sludge and septic systems.

Aquifers

An aquifer is a layer of porous substrate that contains and transmits groundwater.

When water can flow directly between the surface and the saturated zone of an aquifer, the aquifer is unconfined. The deeper parts of unconfined aquifers are usually more saturated since gravity causes water to flow downward.

The upper level of this saturated layer of an unconfined aquifer is called the water table or phreatic surface. Below the water table, where in general all pore spaces are saturated with water, is the phreatic zone.

Substrate with low porosity that permits limited transmission of groundwater is known as an aquitard. An aquiclude is a substrate with porosity that is so low it is virtually impermeable to groundwater.

A confined aquifer is an aquifer that is overlain by a relatively impermeable layer of rock or substrate such as an aquiclude or aquitard. If a confined aquifer follows a downward grade from its recharge zone, groundwater can become pressurized as it flows. This can create artesian wells that flow freely without the need of a pump and rise to a higher elevation than the static water table at the above, unconfined, aquifer.

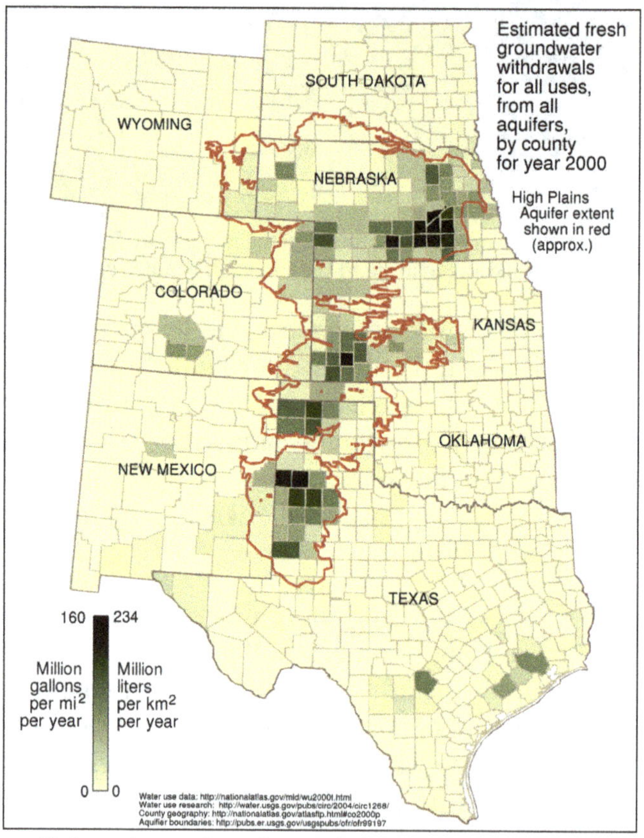

Groundwater withdrawal rates from the Ogallala Aquifer in the Central United States.

The characteristics of aquifers vary with the geology and structure of the substrate and topography in which they occur. In general, the more productive aquifers occur in sedimentary geologic formations. By comparison, weathered and fractured crystalline rocks yield smaller quantities of groundwater in many environments. Unconsolidated to poorly cemented alluvial materials that have accumulated as valley-filling sediments in major river valleys and geologically subsiding structural basins are included among the most productive sources of groundwater.

The high specific heat capacity of water and the insulating effect of soil and rock can mitigate the effects of climate and maintain groundwater at a relatively steady temperature. In some places where groundwater temperatures are maintained by this effect at about 10 °C (50 °F), groundwater can be used for controlling the temperature inside structures at the surface. For example, during hot weather relatively cool groundwater can be pumped through radiators in a home and then returned to the ground in another well. During cold seasons, because it is relatively warm, the water can be used in the same way as a source of heat for heat pumps that is much more efficient than using air.

The volume of groundwater in an aquifer can be estimated by measuring water levels in local wells and by examining geologic records from well-drilling to determine the extent, depth and thickness of water-bearing sediments and rocks. Before an investment is made in production wells, test wells may be drilled to measure the depths at which water is encountered and collect samples of soils, rock and water for laboratory analyses. Pumping tests can be performed in test wells to determine flow characteristics of the aquifer.

Water Cycle

Groundwater makes up about twenty percent of the world's fresh water supply, which is about 0.61% of the entire world's water, including oceans and permanent ice. Global groundwater storage is roughly equal to the total amount of freshwater stored in the snow and ice pack, including the north and south poles. This makes it an important resource that can act as a natural storage that can buffer against shortages of surface water, as in during times of drought.

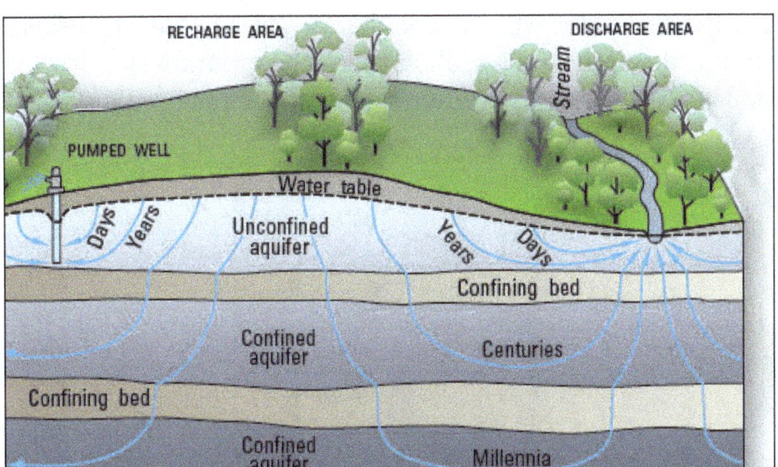

Relative groundwater travel times.

Dzherelo, a common source of drinking water in a Ukrainian village.

Groundwater is naturally replenished by surface water from precipitation, streams, and rivers when this recharge reaches the water table.

Groundwater can be a long-term 'reservoir' of the natural water cycle (with residence times from days to millennia), as opposed to short-term water reservoirs like the atmosphere and fresh surface water (which have residence times from minutes to years). The figure shows how deep groundwater (which is quite distant from the surface recharge) can take a very long time to complete its natural cycle.

The Great Artesian Basin in central and eastern Australia is one of the largest confined aquifer systems in the world, extending for almost 2 million km2. By analysing the trace elements in water sourced from deep underground, hydrogeologists have been able to determine that water extracted from these aquifers can be more than 1 million years old.

By comparing the age of groundwater obtained from different parts of the Great Artesian Basin, hydrogeologists have found it increases in age across the basin. Where water recharges the aquifers along the Eastern Divide, ages are young. As groundwater flows westward across the continent, it increases in age, with the oldest groundwater occurring in the western parts. This means that in order to have travelled almost 1000 km from the source of recharge in 1 million years, the groundwater flowing through the Great Artesian Basin travels at an average rate of about 1 metre per year.

Reflective carpet trapping soil water vapor.

Recent research has demonstrated that evaporation of groundwater can play a significant role in the local water cycle, especially in arid regions. Scientists in Saudi Arabia have proposed plans to recapture and recycle this evaporative moisture for crop irrigation. In the opposite photo, a 50-centimeter-square reflective carpet, made of small adjacent plastic cones, was placed in a plant-free dry desert area for five months, without rain or irrigation. It managed to capture and condense enough ground vapor to bring to life naturally buried seeds underneath it, with a green area of about 10% of the carpet area. It is expected that, if seeds were put down before placing this carpet, a much wider area would become green.

Issues

Certain problems have beset the use of groundwater around the world. Just as river waters have been over-used and polluted in many parts of the world, so too have aquifers. The big difference is that aquifers are out of sight. The other major problem is that water management agencies, when calculating the "sustainable yield" of aquifer and river water, have often counted the same water twice, once in the aquifer, and once in its connected river. This problem, although understood for centuries, has persisted, partly through inertia within government agencies. In Australia, for example, prior to the statutory reforms initiated by the Council of Australian Governments water reform framework in the 1990s, many Australian states managed groundwater and surface water through separate government agencies, an approach beset by rivalry and poor communication.

In general, the time lags inherent in the dynamic response of groundwater to development have been ignored by water management agencies, decades after scientific understanding of the issue was consolidated. In brief, the effects of groundwater overdraft (although undeniably real) may take decades or centuries to manifest themselves. In a classic study in 1982, Bredehoeft and colleagues modeled a situation where groundwater extraction in an intermontane basin withdrew the entire annual recharge, leaving 'nothing' for the natural groundwater-dependent vegetation community. Even when the borefield was situated close to the vegetation, 30% of the original vegetation demand could still be met by the lag inherent in the system after 100 years. By year 500, this had reduced to 0%, signalling complete death of the groundwater-dependent vegetation. The science has been available to make these calculations for decades; however, in general water management agencies have ignored effects that will appear outside the rough timeframe of political elections (3 to 5 years). Marios Sophocleous argued strongly that management agencies must define and use appropriate timeframes in groundwater planning. This will mean calculating groundwater withdrawal permits based on predicted effects decades, sometimes centuries in the future.

As water moves through the landscape, it collects soluble salts, mainly sodium chloride. Where such water enters the atmosphere through evapotranspiration, these salts are left behind. In irrigation districts, poor drainage of soils and surface aquifers can result in water tables' coming to the surface in low-lying areas. Major land degradation problems of soil salinity and waterlogging result, combined with increasing levels of salt in surface waters. As a consequence, major damage has occurred to local economies and environments.

Four important effects are worthy of brief mention. First, flood mitigation schemes, intended to protect infrastructure built on floodplains, have had the unintended consequence of reducing aquifer recharge associated with natural flooding. Second, prolonged depletion of groundwater in extensive aquifers can result in land subsidence, with associated infrastructure damage – as well as, third, saline intrusion. Fourth, draining acid sulphate soils, often found in low-lying coastal plains, can result in acidification and pollution of formerly freshwater and estuarine streams.

Another cause for concern is that groundwater drawdown from over-allocated aquifers has the potential to cause severe damage to both terrestrial and aquatic ecosystems – in some cases very conspicuously but in others quite imperceptibly because of the extended period over which the damage occurs.

Overdraft

Wetlands contrast the arid landscape around Middle Spring, Fish Springs National Wildlife Refuge, Utah.

Groundwater is a highly useful and often abundant resource. However, over-use, over-abstraction or overdraft, can cause major problems to human users and to the environment. The most evident problem (as far as human groundwater use is concerned) is a lowering of the water table beyond the reach of existing wells. As a consequence, wells must be drilled deeper to reach the groundwater; in some places (e.g., California, Texas, and India) the water table has dropped hundreds of feet because of extensive well pumping. In the Punjab region of India, for example, groundwater levels have dropped 10 meters since 1979, and the rate of depletion is accelerating. A lowered water table may, in turn, cause other problems such as groundwater-related subsidence and saltwater intrusion.

Groundwater is also ecologically important. The importance of groundwater to ecosystems is often overlooked, even by freshwater biologists and ecologists. Groundwaters sustain rivers, wetlands, and lakes, as well as subterranean ecosystems within karst or alluvial aquifers.

Not all ecosystems need groundwater, of course. Some terrestrial ecosystems – for example, those of the open deserts and similar arid environments – exist on irregular rainfall and the moisture it delivers to the soil, supplemented by moisture in the air. While there are other terrestrial ecosystems in more hospitable environments where groundwater plays no central role, groundwater is in fact fundamental to many of the world's major ecosystems. Water flows between groundwaters and surface waters. Most rivers, lakes, and wetlands are fed by, and (at other places or times) feed groundwater, to varying degrees. Groundwater feeds soil moisture through percolation, and many terrestrial vegetation communities depend directly on either groundwater or the percolated soil moisture above the aquifer for at least part of each year. Hyporheic zones (the mixing zone of streamwater and groundwater) and riparian zones are examples of ecotones largely or totally dependent on groundwater.

Subsidence

Subsidence occurs when too much water is pumped out from underground, deflating the space below the above-surface, and thus causing the ground to collapse. The result can look like craters on plots of land. This occurs because, in its natural equilibrium state, the hydraulic pressure of groundwater in the pore spaces of the aquifer and the aquitard supports some of the weight of the overlying sediments. When groundwater is removed from aquifers by excessive pumping, pore pressures in the aquifer drop and compression of the aquifer may occur. This compression may be partially recoverable if pressures rebound, but much of it is not. When the aquifer gets compressed, it may cause land subsidence, a drop in the ground surface. The city of New Orleans, Louisiana is actually below sea level today, and its subsidence is partly caused by removal of groundwater from the various aquifer/aquitard systems beneath it. In the first half of the 20th century, the San Joaquin Valley experienced significant subsidence, in some places up to 8.5 metres (28 feet) due to groundwater removal. Cities on river deltas, including Venice in Italy, and Bangkok in Thailand, have experienced surface subsidence; Mexico City, built on a former lake bed, has experienced rates of subsidence of up to 40 cm (1'3") per year.

Seawater Intrusion

In general, in very humid or undeveloped regions, the shape of the water table mimics the slope of the surface. The recharge zone of an aquifer near the seacoast is likely to be inland, often at considerable distance. In these coastal areas, a lowered water table may induce sea water to reverse the flow toward the land. Sea water moving inland is called a saltwater intrusion. In alternative fashion, salt from mineral beds may leach into the groundwater of its own accord.

Pollution

Iron (III) oxide staining (after water capillary rise in a wall) caused by oxidation of dissolved iron (II) and its subsequent precipitation, from an unconfined aquifer in karst topography. Perth, Western Australia.

Polluted groundwater is less visible, but more difficult to clean up, than pollution in rivers and lakes. Groundwater pollution most often results from improper disposal of wastes on land. Major sources include industrial and household chemicals and garbage landfills, industrial waste lagoons, tailings and process wastewater from mines, oil field brine pits, leaking underground oil storage tanks and pipelines, sewage sludge and septic systems. Polluted groundwater is mapped by sampling soils and groundwater near suspected or known sources of pollution, to determine the extent of the pollution, and to aid in the design of groundwater remediation systems. Preventing groundwater pollution near potential sources such as landfills requires lining the bottom of a landfill with watertight materials, collecting any leachate with drains, and keeping rainwater off any potential contaminants, along with regular monitoring of nearby groundwater to verify that contaminants have not leaked into the groundwater.

Groundwater pollution, from pollutants released to the ground that can work their way down into groundwater, can create a contaminant plume within an aquifer. Pollution can occur from landfills, naturally occurring arsenic, on-site sanitation systems or other point sources, such as petrol stations with leaking underground storage tanks, or leaking sewers.

Movement of water and dispersion within the aquifer spreads the pollutant over a wider area, its advancing boundary often called a plume edge, which can then intersect with groundwater wells or daylight into surface water such as seeps and springs, making the water supplies unsafe for humans and wildlife. Different mechanism have influence on the transport of pollutants, e.g. diffusion, adsorption, precipitation, decay, in the groundwater. The interaction of groundwater contamination with surface waters is analyzed by use of hydrology transport models.

The danger of pollution of municipal supplies is minimized by locating wells in areas of deep groundwater and impermeable soils, and careful testing and monitoring of the aquifer and nearby potential pollution sources.

Arsenic and Fluoride

Around one-third of the world's population drinks water from groundwater resources. Of this, about 10 percent, approximately 300 million people, obtains water from groundwater resources that are heavily polluted with arsenic or fluoride. These trace elements derive mainly from natural sources by leaching from rock and sediments.

New Method of Identifying Substances that are Hazardous to Health

In 2008, the Swiss Aquatic Research Institute, Eawag, presented a new method by which hazard maps could be produced for geogenic toxic substances in groundwater. This provides an efficient way of determining which wells should be tested.

In 2016, the research group made its knowledge freely available on the Groundwater Assessment Platform GAP. This offers specialists worldwide the possibility of uploading their own measurement data, visually displaying them and producing risk maps for areas of their choice. GAP also serves as a knowledge-sharing forum for enabling further development of methods for removing toxic substances from water.

WELLS

A well is an excavation or structure created in the ground by digging, driving, or drilling to access liquid resources, usually water. The oldest and most common kind of well is a water well, to access groundwater in underground aquifers. The well water is drawn by a pump, or using containers, such as buckets, that are raised mechanically or by hand. Wells were first constructed at least eight thousand years ago and historically vary in construction from a simple scoop in the sediment of a dry watercourse to the qanats of Iran, and the stepwells and sakiehs of India. Placing a lining in the well shaft helps create stability, and linings of wood or wickerwork date back at least as far as the Iron Age.

Wells have traditionally been sunk by hand digging, as is the case in rural areas of the developing world. These wells are inexpensive and low-tech as they use mostly manual labour, and the structure can be lined with brick or stone as the excavation proceeds. A more modern method called caissoning uses pre-cast reinforced concrete well rings that are lowered into the hole. Driven wells can be created in unconsolidated material with a well hole structure, which consists of a hardened drive point and a screen of perforated pipe, after which a pump is installed to collect the water. Deeper wells can be excavated by hand drilling methods or machine drilling, using a bit in a borehole. Drilled wells are usually cased with a factory-made pipe composed of steel or plastic. Drilled wells can access water at much greater depths than dug wells.

Two broad classes of well are shallow or unconfined wells completed within the uppermost saturated aquifer at that location, and deep or confined wells, sunk through an impermeable stratum into an aquifer beneath. A collector well can be constructed adjacent to a freshwater lake or stream with water percolating through the intervening material. The site of a well can be selected by a hydrogeologist, or groundwater surveyor. Water may be pumped or hand drawn. Impurities from the surface can easily reach shallow sources and contamination of the supply by pathogens or chemical contaminants needs to be avoided. Well water typically contains more minerals in solution than surface water and may require treatment before being potable. Soil salination can occur as the water table falls and the surrounding soil begins to dry out. Another environmental problem is the potential for methane to seep into the water.

Camel drawing water from a well, Djerba island, Tunisia.

A Chinese ceramic model of a well with a water pulley system, excavated from a tomb of the Han Dynasty period.

Wood-lined wells are known from the early Neolithic Linear Pottery culture, for example in Kückhoven (an outlying centre of Erkelenz), dated 5090 BC and Eythra, dated 5200 BC in Schletz (an outlying centre of Asparn an der Zaya) in Austria.

Some of the earliest evidence of water wells are located in China. The neolithic Chinese discovered and made extensive use of deep drilled groundwater for drinking.

The Chinese text The Book of Changes, originally a divination text of the Western Zhou dynasty, contains an entry describing how the ancient Chinese maintained their wells and protected their sources of water. Archaeological evidence and old Chinese documents reveal that the prehistoric and ancient Chinese had the aptitude and skills for digging deep water wells for drinking water as early as 6000 to 7000 years ago. A well excavated at the Hemedu excavation site was believed to have been built during the neolithic era. The well was cased by four rows of logs with a square frame attached to them at the top of the well. 60 additional tile wells southwest of Beijing are also believed to have been built around 600 BC for drinking and irrigation.

In Egypt, shadoofs and sakiehs are used. When compared to each other however, the Sakkieh is much more efficient, as it can bring up water from a depth of 10 metres (versus the 3 metres of the shadoof). The Sakieh is the Egyptian version of the Noria. Some of the world's oldest known wells, located in Cyprus, date to 7000-8500 BC. Two wells from the Neolithic period, around 6500 BC, have been discovered in Israel. One is in Atlit, on the northern coast of Israel, and the other is the Jezreel Valley.

Wells for other purposes came along much later, historically. The first recorded salt well was dug in the Sichuan province of China around 2,250 years ago. This was the first time that ancient water well technology was applied successfully for the exploitation of salt, and marked the beginning of Sichuan's salt drilling industry. The earliest known oil wells were also drilled in China, in 347 CE. These wells had depths of up to about 240 metres (790 ft) and were drilled using bits attached to bamboo poles. The oil was burned to evaporate brine and produce salt. By the 10th century, extensive bamboo pipelines connected oil wells with salt springs. The ancient records of China and Japan are said to contain many allusions to the use of natural gas for lighting and heating. Petroleum was known as Burning water in Japan in the 7th century.

Types

Water well near Simaisma, eastern Qatar.

Leather bucket used for the water well.

Well, Historical Village, Bhaini Sahib, Ludhiana, Punjab, India.

Dug wells

Until recent centuries, all artificial wells were pumpless hand-dug wells of varying degrees of sophistication, and they remain a very important source of potable water in some rural developing areas where they are routinely dug and used today. Their indispensability has produced a number of literary references, literal and figurative, to them, including the reference to the incident of Jesus meeting a woman at Jacob's well in the bible and the "Ding Dong Bell" nursery rhyme about a cat in a well.

Hand-dug wells are excavations with diameters large enough to accommodate one or more people with shovels digging down to below the water table. The excavation is braced horizontally to avoid landslide or erosion endangering the people digging. They can be lined with laid stones or brick; extending this lining upwards above the ground surface to form a wall around the well serves to reduce both contamination and injuries by falling into the well. A more modern method called caissoning uses reinforced concrete or plain concrete pre-cast well rings that are lowered into the hole. A well-digging team digs under a cutting ring and the well column slowly sinks into the aquifer, whilst protecting the team from collapse of the well bore.

Hand-dug wells are inexpensive and low tech (compared to drilling) as they use mostly manual labour to access groundwater in rural locations in developing countries. They may be built with a high degree of community participation, or by local entrepreneurs who specialize in hand-dug wells. They have been successfully excavated to 60 metres (200 ft). They have low operational and maintenance costs, in part because water can be extracted by hand bailing, without a pump. The water is often coming from an aquifer or groundwater, and can be easily deepened, which may be necessary if the ground water level drops, by telescoping the lining further down into the aquifer. The yield of existing hand dug wells may be improved by deepening or introducing vertical tunnels or perforated pipes.

Drawbacks to hand-dug wells are numerous. It can be impractical to hand dig wells in areas where hard rock is present, and they can be time-consuming to dig and line even in favourable areas. Because they exploit shallow aquifers, the well may be susceptible to yield fluctuations and possible contamination from surface water, including sewage. Hand dug well construction generally requires the use of a well trained construction team, and the capital investment for equipment such as concrete ring moulds, heavy lifting equipment, well shaft formwork, motorized de-watering pumps, and fuel can be large for people in developing countries. Construction of hand dug wells can be dangerous due to collapse of the well bore, falling objects and asphyxiation, including from dewatering pump exhaust fumes.

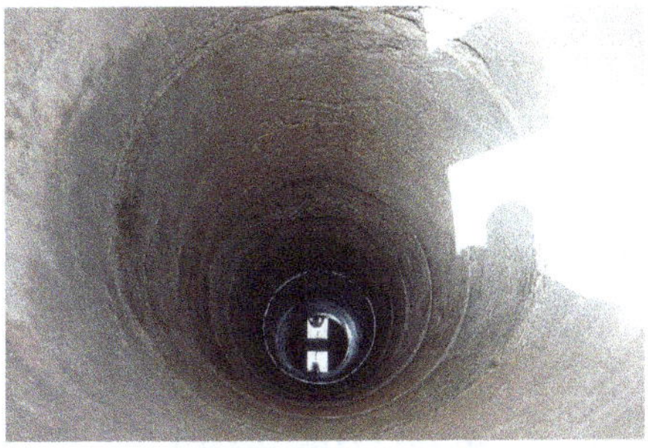

View into a hand-dug well cased with concrete rings; Ouelessebougou, Mali.

A dug well in a village in Kerala, India.

Woodingdean well, hand-dug between 1858 and 1862, is claimed to be the world's deepest hand-dug well at 392 metres (1,285 ft). The Big Well in Greensburg, Kansas is billed as the world's largest hand-dug well, at 109 feet (33 m) deep and 32 feet (9.8 m) in diameter. However, the Well of Joseph in the Cairo Citadel at 280 feet (85 m) deep and the Pozzo di S. Patrizio (St. Patrick's Well) built in 1527 in Orvieto, Italy, at 61 metres (200 ft) deep by 13 metres (43 ft) wide are both larger by volume.

Driven Wells

Driven wells may be very simply created in unconsolidated material with a well hole structure, which consists of a hardened drive point and a screen (perforated pipe). The point is simply hammered into the ground, usually with a tripod and driver, with pipe sections added as needed. A driver is a weighted pipe that slides over the pipe being driven and is repeatedly dropped on it. When groundwater is encountered, the well is washed of sediment and a pump installed.

Drilled Wells

Drilled wells are typically created using either top-head rotary style, table rotary, or cable tool drilling machines, all of which use drilling stems that are turned to create a cutting action in the formation, hence the term drilling.

Drilled wells can be excavated by simple hand drilling methods (augering, sludging, jetting, driving, hand percussion) or machine drilling (rotary, percussion, down the hole hammer). Deeprock rotary drilling method is most common. Rotary can be used in 90% of formation types.

Drilled wells can get water from a much deeper level than dug wells can—often down to several hundred metres.

Drilled wells with electric pumps are used throughout the world, typically in rural or sparsely populated areas, though many urban areas are supplied partly by municipal wells. Most shallow well drilling machines are mounted on large trucks, trailers, or tracked vehicle carriages. Water wells typically range from 3 to 18 metres (10–60 ft) deep, but in some areas can go deeper than 900 metres (3,000 ft).

Cable tool water well drilling rig in Kimball, West Virginia, USA.

Water well drilling in Ein Hemed, near Jerusalem.

Rotary drilling machines use a segmented steel drilling string, typically made up of 6 m (20 ft) sections of galvanized steel tubing that are threaded together, with a bit or other drilling device at the bottom end. Some rotary drilling machines are designed to install (by driving or drilling) a steel casing into the well in conjunction with the drilling of the actual bore hole. Air and/or water is used as a circulation fluid to displace cuttings and cool bits during the drilling. Another form of rotary-style drilling, termed mud rotary, makes use of a specially made mud, or drilling fluid, which is constantly being altered during the drill so that it can consistently create enough hydraulic pressure to hold the side walls of the bore hole open, regardless of the presence of a casing in the well. Typically, boreholes drilled into solid rock are not cased until after the drilling process is completed, regardless of the machinery used.

The oldest form of drilling machinery is the cable tool, still used today. Specifically designed to raise and lower a bit into the bore hole, the spudding of the drill causes the bit to be raised and dropped onto the bottom of the hole, and the design of the cable causes the bit to twist at approximately 1⁄4 revolution per drop, thereby creating a drilling action. Unlike rotary drilling, cable tool drilling requires the drilling action to be stopped so that the bore hole can be bailed or emptied of drilled cuttings.

Drilled wells are usually cased with a factory-made pipe, typically steel (in air rotary or cable tool drilling) or plastic/PVC (in mud rotary wells, also present in wells drilled into solid rock). The casing is constructed by welding, either chemically or thermally, segments of casing together. If the casing is installed during the drilling, most drills will drive the casing into the ground as the bore hole advances, while some newer machines will actually allow for the casing to be rotated and drilled into the formation in a similar manner as the bit advancing just below. PVC or plastic is typically welded and then lowered into the drilled well, vertically stacked with their ends nested and either glued or splined together. The sections of casing are usually 6 metres (20 ft) or more in length, and 6 to 12 in (15 to 30 cm) in diameter, depending on the intended use of the well and local groundwater conditions.

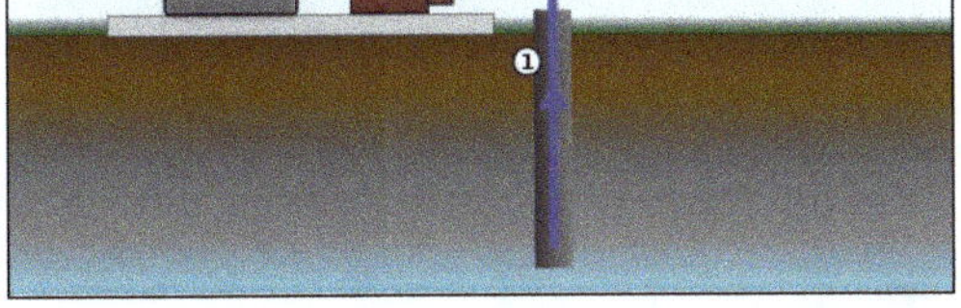

An automated water well system powered by a jet-pump.

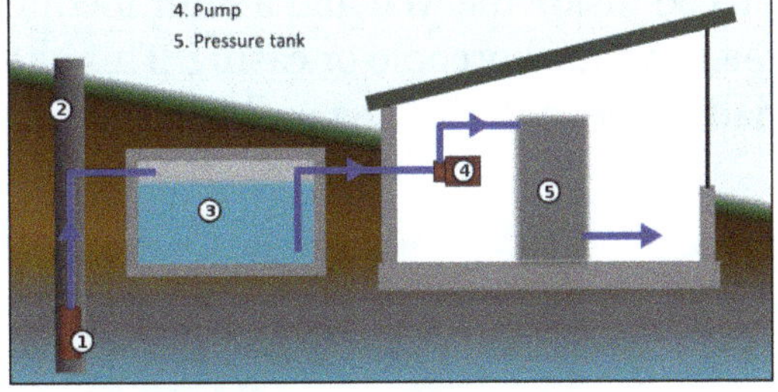

A water well system with a cistern.

Surface contamination of wells in the United States is typically controlled by the use of a surface seal. A large hole is drilled to a predetermined depth or to a confining formation (clay or bedrock, for example), and then a smaller hole for the well is completed from that point forward. The well is typically cased from the surface down into the smaller hole with a casing that is the same diameter as that hole. The annular space between the large bore hole and the smaller casing is filled with bentonite clay, concrete, or other sealant material. This creates an impermeable seal from the surface to the next confining layer that keeps contaminants from traveling down the outer sidewalls of the casing or borehole and into the aquifer. In addition, wells are typically capped with either an engineered well cap or seal that vents air through a screen into the well, but keeps insects, small animals, and unauthorized persons from accessing the well.

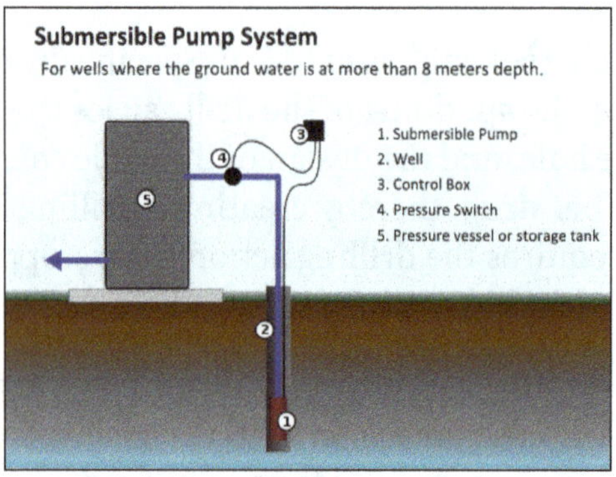

An automated water well system powered by a submersible pump.

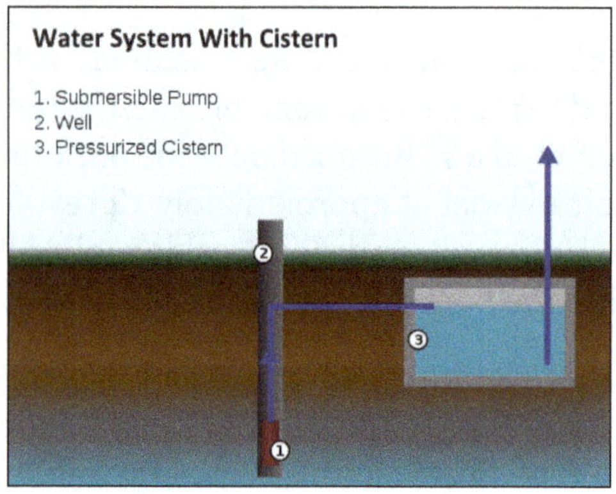

A water well system with a pressurized cistern.

At the bottom of wells, based on formation, a screening device, filter pack, slotted casing, or open bore hole is left to allow the flow of water into the well. Constructed screens are typically used in unconsolidated formations (sands, gravels, etc.), allowing water and a percentage of the formation to pass through the screen. Allowing some material to pass through creates a large area filter out of the rest of the formation, as the amount of material present to pass into the well slowly decreases and is removed from the well. Rock wells are typically cased with a PVC liner/casing and screen or slotted casing at the bottom, this is mostly present just to keep rocks from entering the pump assembly. Some wells utilize a filter pack method, where an undersized screen or slotted casing is placed inside the well and a filter medium is packed around the screen, between the screen and the borehole or casing. This allows the water to be filtered of unwanted materials before entering the well and pumping zone.

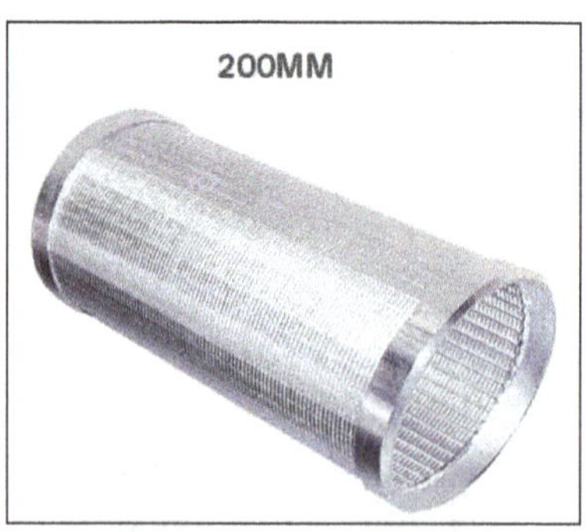

A section of a stainless steel screen well.

Classification

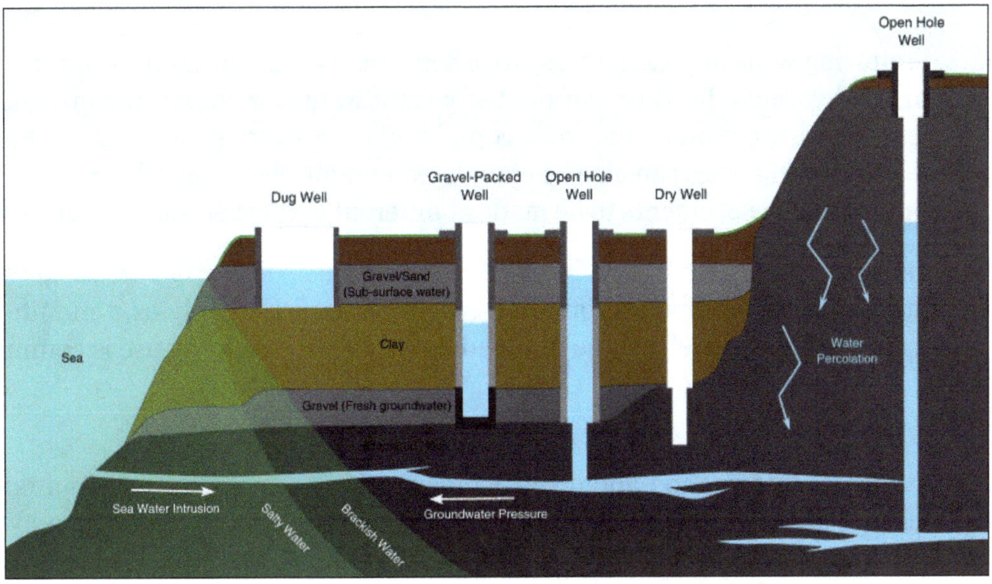

Water Well Types

There are two broad classes of drilled-well types, based on the type of aquifer the well is in:

- Shallow or unconfined wells are completed in the uppermost saturated aquifer at that location (the upper unconfined aquifer).
- Deep or confined wells are sunk through an impermeable stratum into an aquifer that is sandwiched between two impermeable strata (aquitards or aquicludes). The majority of deep aquifers are classified as artesian because the hydraulic head in a confined well is higher than the level of the top of the aquifer. If the hydraulic head in a confined well is higher than the land surface it is a "flowing" artesian well (named after Artois in France).

A special type of water well may be constructed adjacent to freshwater lakes or streams. Commonly called a collector well but sometimes referred to by the trade name Ranney well or Ranney collector, this type of well involves sinking a caisson vertically below the top of the aquifer and then advancing lateral collectors out of the caisson and beneath the surface water body. Pumping from within the caisson induces infiltration of water from the surface water body into the aquifer, where it is collected by the collector well laterals and conveyed into the caisson where it can be pumped to the ground surface.

Two additional broad classes of well types may be distinguished, based on the use of the well:

- Production or pumping wells, are large diameter (greater than 15 cm in diameter) cased (metal, plastic, or concrete) water wells, constructed for extracting water from the aquifer by a pump (if the well is not artesian).

- Monitoring wells or piezometers, are often smaller diameter wells used to monitor the hydraulic head or sample the groundwater for chemical constituents. Piezometers are monitoring wells completed over a very short section of aquifer. Monitoring wells can also be completed at multiple levels, allowing discrete samples or measurements to be made at different vertical elevations at the same map location.

A well constructed for pumping groundwater can be used passively as a monitoring well and a small diameter well can be pumped, but this distinction by use is common.

Siting

Before excavation, information about the geology, water table depth, seasonal fluctuations, recharge area and rate must be found. This work is typically done by a hydrogeologist, or a groundwater surveyor using a variety of tools including electro-seismic surveying, any available information from nearby wells, geologic maps, and sometimes geophysical imaging.

Contamination

Waterborne diseases can be spread via a well which is contaminated with fecal pathogens from pit latrines.

Shallow pumping wells can often supply drinking water at a very low cost. However, impurities from the surface easily reach shallow sources, which leads to a greater risk of contamination for these wells compared to deeper wells. Contaminated wells can lead to the spread of various waterborne diseases. Dug and driven wells are relatively easy to contaminate; for instance, most dug wells are unreliable in the majority of the United States.

Man cleaning a well in Yaoundé, Cameroon.

Hand pump to pump water from a well in a village near Chennai in India, where the well water might be polluted by nearby pit latrines.

Pathogens

Most of the bacteria, viruses, parasites, and fungi that contaminate well water comes from fecal material from humans and other animals, for example from on-site sanitation systems (such as pit latrines and septic tanks). Common bacterial contaminants include E. coli, Salmonella, Shigella, and Campylobacter jejuni. Common viral contaminants include norovirus, sapovirus, rotavirus, enteroviruses, and hepatitis A and E. Parasites include Giardia lamblia, Cryptosporidium, Cyclospora cayetanensis, and microsporidia.

Chemical Contamination

Chemical contamination is a common problem with groundwater. Nitrates from sewage, sewage sludge or fertilizer are a particular problem for babies and young children. Pollutant chemicals include pesticides and volatile organic compounds from gasoline, dry-cleaning, the fuel additive methyl tert-butyl ether (MTBE), and perchlorate from rocket fuel, airbag inflators, and other artificial and natural sources.

Several minerals are also contaminants, including lead leached from brass fittings or old lead pipes, chromium VI from electroplating and other sources, naturally occurring arsenic, radon, and uranium—all of which can cause cancer—and naturally occurring fluoride, which is desirable in low quantities to prevent tooth decay, but can cause dental fluorosis in higher concentrations.

Some chemicals are commonly present in water wells at levels that are not toxic, but can cause other problems. Calcium and magnesium cause what is known as hard water, which can precipitate and clog pipes or burn out water heaters. Iron and manganese can appear as dark flecks that stain clothing and plumbing, and can promote the growth of iron and manganese bacteria that can form slimy black colonies that clog pipes.

Prevention

The quality of the well water can be significantly increased by lining the well, sealing the well head, fitting a self-priming hand pump, constructing an apron, ensuring the area is kept clean and free from stagnant water and animals, moving sources of contamination (pit latrines, garbage pits, on-site sewer systems) and carrying out hygiene education. The well should be cleaned with 1% chlorine solution after construction and periodically every 6 months.

Well holes should be covered to prevent loose debris, animals, animal excrement, and wind-blown foreign matter from falling into the hole and decomposing. The cover should be able to be in place at all times, including when drawing water from the well. A suspended roof over an open hole helps to some degree, but ideally the cover should be tight fitting and fully enclosing, with only a screened air vent.

Minimum distances and soil percolation requirements between sewage disposal sites and water wells need to be observed. Rules regarding the design and installation of private and municipal septic systems take all these factors into account so that nearby drinking water sources are protected.

Education of the general population in society also plays an important role in protecting drinking water.

Mitigation

Cleanup of contaminated groundwater tends to be very costly. Effective remediation of groundwater is generally very difficult. Contamination of groundwater from surface and subsurface sources can usually be dramatically reduced by correctly centering the casing during construction and filling the casing annulus with an appropriate sealing material. The sealing material (grout) should be placed from immediately above the production zone back to surface, because, in the absence of a correctly constructed casing seal, contaminated fluid can travel into the well through the casing annulus. Centering devices are important (usually 1 per length of casing or at maximum intervals of 9 m) to ensure that the grouted annular space is of even thickness. Upon the construction of a new test well, it is considered best practice to invest in a complete battery of chemical and biological tests on the well water in question. Point-of-use treatment is available for individual properties and treatment plants are often constructed for municipal water supplies that suffer from contamination. Most of these treatment methods involve the filtration of the contaminants of concern, and additional protection may be garnered by installing well-casing screens only at depths where contamination is not present.

Well water for personal use is often filtered with reverse osmosis water processors; this process can remove very small particles. A simple, effective way of killing microorganisms is to bring the water to a full boil for one to three minutes, depending on location. A household well contaminated by microorganisms can initially be treated by shock chlorination using bleach, generating concentrations hundreds of times greater than found in community water systems; however, this will not fix any structural problems that led to the contamination and generally requires some expertise and testing for effective application.

After the filtration process, it is common to implement an Ultraviolet (UV) system to kill pathogens in the water. UV light effects the DNA of the pathogen by UV-C photons breaking through the cell wall. UV disinfection has been gaining popularity in the past decades as it is a chemical free method of water treatment.

Environmental Problems

A risk with the placement of water wells is soil salination which occurs when the water table of the soil begins to drop and salt begins to accumulate as the soil begins to dry out. Another environmental problem that is very prevalent in water well drilling is the potential for methane to seep through.

Soil Salination

The potential for soil salination is a large risk when choosing the placement of water wells. Soil salination is caused when the water table of the soil drops over time and salt begins to accumulate. In turn, the increased amount of salt begins to dry the soil out. This is a very detrimental problem because the increased level of salt in the soil can result in the degradation of soil and can be very harmful to vegetation.

Methane

Methane, an asphyxiant, is a chemical compound that is the main component of natural gas. When methane is introduced into a confined space, it displaces oxygen, reducing oxygen concentration to a level low enough to pose a threat to humans and other aerobic organisms but still high enough for a risk of spontaneous or externally caused explosion. This potential for explosion is what poses such a danger in regards to the drilling and placement of water wells.

Low levels of methane in drinking water are not considered toxic. When methane seeps into a water supply, it is commonly referred to as "methane migration". This can be caused by old natural gas wells near water well systems becoming abandoned and no longer monitored.

Lately, however, the described wells/pumps are no longer very efficient and can be replaced by either handpumps or treadle pumps. Another alternative is the use of self-dug wells, electrical deep-well pumps (for higher depths). Appropriate technology organizations as Practical Action are now supplying information on how to build/set-up (DIY) handpumps and treadle pumps in practice.

WATER CYCLE

Water cycle, also called hydrologic cycle, is a cycle that involves the continuous circulation of water in the Earth-atmosphere system. Of the many processes involved in the water cycle, the most important are evaporation, transpiration, condensation, precipitation, and runoff. Although the total amount of water within the cycle remains essentially constant, its distribution among the various processes is continually changing.

Evaporation, one of the major processes in the cycle, is the transfer of water from the surface of the Earth to the atmosphere. By evaporation, water in the liquid state is transferred to the gaseous, or vapour, state. This transfer occurs when some molecules in a water mass have attained sufficient kinetic energy to eject themselves from the water surface. The main factors affecting evaporation are temperature, humidity, wind speed, and solar radiation. The direct measurement of evaporation, though desirable, is difficult and possible only at point locations. The principal source of water vapour is the oceans, but evaporation also occurs in soils, snow, and ice. Evaporation from snow and ice, the direct conversion from solid to vapour, is known as sublimation. Transpiration is the evaporation of water through minute pores, or stomata, in the leaves of plants. For practical purposes, transpiration and the evaporation from all water, soils, snow, ice, vegetation, and other surfaces are lumped together and called evapotranspiration, or total evaporation.

Water vapour is the primary form of atmospheric moisture. Although its storage in the atmosphere is comparatively small, water vapour is extremely important in forming the moisture supply for dew, frost, fog, clouds, and precipitation. Practically all water vapour in the atmosphere is confined to the troposphere (the region below 6 to 8 miles [10 to 13 km] altitude).

The transition process from the vapour state to the liquid state is called condensation. Condensation may take place as soon as the air contains more water vapour than it can receive from a free water surface through evaporation at the prevailing temperature.

This condition occurs as the consequence of either cooling or the mixing of air masses of different temperatures. By condensation, water vapour in the atmosphere is released to form precipitation.

Precipitation that falls to the Earth is distributed in four main ways: some is returned to the atmosphere by evaporation, some may be intercepted by vegetation and then evaporated from the surface of leaves, some percolates into the soil by infiltration, and the remainder flows directly as surface runoff into the sea. Some of the infiltrated precipitation may later percolate into streams as groundwater runoff. Direct measurement of runoff is made by stream gauges and plotted against time on hydrographs.

Most groundwater is derived from precipitation that has percolated through the soil. Groundwater flow rates, compared with those of surface water, are very slow and variable, ranging from a few millimetres to a few metres a day. Groundwater movement is studied by tracer techniques and remote sensing.

Ice also plays a role in the water cycle. Ice and snow on the Earth's surface occur in various forms such as frost, sea ice, and glacier ice. When soil moisture freezes, ice also occurs beneath the Earth's surface, forming permafrost in tundra climates. About 18,000 years ago glaciers and ice caps covered approximately one-third of the Earth's land surface. Today about 12 percent of the land surface remains covered by ice masses.

FOSSIL WATER

Fossil water or paleowater is an ancient body of water that has been contained in some undisturbed space, typically groundwater in an aquifer, for millennia. Other types of fossil water can include subglacial lakes, such as Antarctica's Lake Vostok, and even ancient water on other planets.

UNESCO defines fossil groundwater as:

> Water that infiltrated usually millennia ago and often under climatic conditions different from the present, and that has been stored underground since that time.
>
> —*Dr Karin Kemper*

Determining the time since water infiltrated usually involves analyzing isotopic signatures. Determining "fossil" status—whether or not that particular water has occupied that particular space since the distant past—involves modeling the flow, recharge, and losses of aquifers, which can involve significant uncertainty. Some aquifers are hundreds of meters deep and underlie vast areas of land. Research techniques in the field are developing quickly and the scientific knowledge base is growing. In the cases of many aquifers, research is lacking or disputed as to the age of the water and the behavior of the water inside the aquifer.

Renewability

Large, prolific aquifers (notably the Nubian Sandstone Aquifer System and the Ogallala Aquifer) containing fossil water are of significant socio-economic value. Fossil water is extracted from these aquifers for many human purposes, notably, agriculture, industry, and consumption. In arid regions, some aquifers containing available and usable water receive little to no significant recharge, effectively making groundwater in those aquifers a non-renewable resource. Extraction rates greater than recharge rates result in lowering of the water table and can lead to groundwater depletion. Extraction of non-renewable groundwater resources is referred to as groundwater "mining" because of their finite nature.

In other regions that do receive significant precipitation and where recharge does occur, fossil water may be extracted and eventually replaced by younger water, in which case the water in those aquifers could be considered a renewable resource.

General Geology

Aquifers are typically composed of semi-porous rock or unconsolidated material whose pore space has been filled with water. In the relatively rare cases of confined aquifers, an impermeable geologic layer (e.g. clay or calcrete) encloses an aquifer, isolating the water within, sometimes for millennia. More commonly, fossil water is found in arid or semi-arid regions where the climate was significantly more humid in recent geologic history. In some semi-arid regions, the majority of precipitation evaporates before it can infiltrate and result in any significant aquifer recharge.

Most fossil groundwater has been estimated to have originally infiltrated within the Holocene and Pleistocene (10,000–40,000 years ago). Some fossil groundwater is associated with the melting of ice in the time since the last glacial maximum. Dating of groundwater relies on measuring concentrations of certain stable isotopes, including 3 H (tritium) and 18 O ("heavy" oxygen), and comparing values with known concentrations of the geologic past.

Fossil water can potentially dissolve and absorb a number of ions from its host rock. Salinity in groundwater can be higher than seawater. In some cases, some form of treatment is required to make these waters suitable for human use. Saline fossil aquifers can also store significant quantities of oil and natural gas.

Notable Bodies of Fossil Water

Ogallala Aquifer

The Ogallala or High Plains Aquifer sits under 450,000 km^2 of 8 states of the United States of America. It is one of the largest freshwater deposits in the world. The aquifer is composed of unconsolidated alluvial deposits. Groundwater in this aquifer has been dated to have been deposited in the humid time following the last glacial maximum. In much of the aquifer's area, an impermeable layer of calcrete prevents precipitation from infiltrating. In other regions of the aquifer, some relatively small rates of recharge have been measured.

The aquifer supplies water for the many people who live above it and for widespread agricultural uses. In many areas, the water table has dropped drastically due to heavy extraction. Depletion rates are not stabilizing, in fact, they have been increasing in recent decades.

Transport of pipe segments for the Great Manmade River in Libya: a network of pipes that supplies water from the Nubian Sandstone Aquifer System.

Nubian Sandstone Aquifer System

The Nubian Sandstone Aquifer System is located in northeastern Africa, under the nations of Sudan, Libya, Egypt, and Chad, covering about 2,000,000 km^2. It is largely composed of many hydraulically interconnected sandstone aquifers. Some parts of the system are considered to be confined, if somewhat leaky, due to impermeable layers such as marine shales. The water was deposited between 4,000 and 20,000 years ago, varying by specific locality.

The water in the Nubian Sandstone Aquifer System is of high importance to the people living above it, and has been for millennia. In modern times, as demand increases, avoiding depletion and conflict will depend on careful cross-boundary monitoring and planning. Libya and Egypt are currently planning development projects to withdraw significant amounts of the aquifer's fossil water for use. Other fossil aquifers have been identified throughout Northern Africa.

Kalahari Desert Aquifers

The Kalahari Desert in central southern Africa (Botswana, Namibia, and South Africa). Geology of the area includes significant karst formations. Most of the precipitation in the region evaporates before it can contribute to significant recharge of the aquifers below. Whether or not the region's aquifers receive any significant recharge has long been the subject of debate and research. In the northern region of the Kalahari, a deep aquifer in Cave sandstone was found to have isotopic signatures that suggested it had been confined with little to no leakage for long periods of time.

References

- Rasler, Karen A.; Thompson, W. R. (2006). "Contested Territory, Strategic Rivalries, and Conflict Escalation". International Studies Quarterly. 50 (1): 145–168. doi:10.1111/j.1468-2478.2006.00396.x
- Water, science: britannica.com, Retrieved 28 June, 2020
- "Information Sheet on Ramsar Wetlands (RIS)". Ramsar Convention on Wetlands. Archived from the original on March 4, 2009. Retrieved 2 March, 2013
- Surface-water, energy-government-and-defense-magazines, environment: encyclopedia.com, Retrieved 29 July, 2020
- Davidson, N.C. (2014). "How much wetland has the world lost? Long-term and recent trends in global wetland area". Marine and Freshwater Research. 65 (10): 934–941. Doi:10.1071/MF14173
- Water-cycle, science: britannica.com, Retrieved 30 August, 2020

2

Water Scarcity

The lack of freshwater resources to fulfil the demands of water usage in a region is known as water scarcity. The major causes of water scarcity are water pollution and overpopulation. This chapter has been carefully written to provide an easy understanding of the subject of water scarcity.

Whether it's from drought or lack of access, more than a billion people around the world don't have enough clean water.

Acute drought conditions and dwindling natural water resources are focusing more attention on what continues to be a worldwide problem: A lack of access to fresh, potable water.

Water scarcity can be defined as a lack of sufficient water, or not having access to safe water supplies.

Water is a pressing need in many areas of the world. That scarcity is spreading as water is needed to grow and process food, create energy, and serve industry for a continually growing population. Climate change is a key contributing factor.

Clean water is an essential ingredient of a healthy human life, but 1.2 billion people lack access to water, according to recent estimates from the International Water Management Institute. By 2025, two-thirds of the world's population may be facing water shortages, according to the World Wildlife Federation. Available freshwater supplies worldwide continue to dwindle. By 2030, water demand is forecast to increase by 40%, according to Even Kuross, a management consultant based in Oslo, writing in Fair Observer. The world population is expected to reach 9 billion, placing pressure on water supplies.

Physical Water Scarcity

Physical water scarcity occurs when there isn't enough water to meet demand. Roughly 20% of the world's population now lives in physical water scarcity, which The World's Water: Volume 8 defines as areas in which water withdrawals exceed 75% of river flows. Another 500 million live in areas "approaching physical scarcity." This could be the result of dry or arid local conditions, but distribution also plays a role. The Water Project points out the Colorado River basin as a prime example "of a seemingly abundant source of water being overused and over managed, leading to very serious physical water scarcity downstream."

A child helps his father carry water containers as they fetch drinking water in Bangladesh. Because the area is surrounded by saline water, scarcity of drinking water is a major problem.

Water Economics

There is another equally challenging source of water scarcity: economic factors. The Water Project explains:

> In the developing world, finding a reliable source of safe water is often time consuming and expensive. This is known as economic scarcity. Water can be found it simply requires more resources to do it. Economic water scarcity is by far the most disturbing form of water scarcity because it is almost entirely a lack of compassion and good governance that allows the condition to persist. Economic water scarcity exists when a population does not have the necessary monetary means to utilize an adequate source of water.

Economic water scarcity is predominant throughout Africa, particularly in sub-Saharan Africa. An estimated 1.6 billion people around the world live in areas of economic water scarcity, with 780 million people living in areas with no basic water services. Compounding the lack of infrastructure investment are political and ethnic conflicts, which continue to increase and intensify worldwide as water becomes more and more scarce, writes Brian Richter in the book Chasing Water: A Guide for Moving from Scarcity to Sustainability. Inadequate water supplies can also contribute to political and economic instability.

Population Pressure

The Worldwatch Institute's Supriya Kumar told Voice of America that water scarcity will continue worsening worldwide as the global population continues to grow:

> Over 1.2 billion are basically living in areas of physical water scarcity. And

almost 1.6 billion face economic water shortage. And these are really extreme numbers. And as our population continues to grow there's just going to be more problems. And we're going to really have to face drastic measures in order to make sure the people have access to water.

In the biennial compendium of freshwater information and data, Gleick writes that one key challenge inherent in quantifying the problem is that data is not gathered reliably or consistently. Some of the latest water use data available is actually 20 or more years old. Without reliable, baseline data, many key issues cannot be adequately addressed by policymakers.

Select Water Footprint Per Capita Data, 1996- 2005					
Nation	Population	Rain Water	Surface and Ground water usage	Freshwater Pollution	Total
	(thousands)	(Cubic meters per person per year)			
Australia	19,320.00	1,853.30	216.30	245.00	2,314.60
Bolivia	8,408.00	3,359.90	62.70	45.30	3,467.90
Congo, Democratic Republic	52,052.00	540.00	5.40	6.60	552.10
Cyprus	790.90	1,682.30	349.30	353.80	2,385.40
Israel	6,134.00	1,790.50	253.30	259.00	2,302.70
Niger	11,272.10	3,411.00	87.10	20.50	3,518.70
Saudi Arabia	21,114.20	1,131.20	447.50	270.60	1,849.30
United Arab Emirates	3,329.80	1,921.20	570.60	644.20	3,136.00
United States	288,958.20	1,968.30	238.90	635.30	2,842.50
World	6,154,564.20	1,015.40	153.30	216.50	1,385.20

Water Scarcity Solutions

There are several available solutions able to effectively address water scarcity, including water reuse, storage, management, conservation, and numerous water treatment technologies such as desalination. Typically, one or more approaches must be adopted in tandem to be effective, whether a water-reliant corporation or a government entity is doing the adopting. The crux of the issue is balancing available supply with demand or consumption. Adding water supply through reuse or desalination, for example, isn't a panacea. Without water management and strategies for adequately addressing ever-increasing demand, the solution is incomplete.

Let's look at a few of these solutions, as well as how and where they are being implemented.

Aquifer Recharging

Groundwater is water that collects below the earth's surface in fissures and crevasses, then moves into aquifers. An aquifer is a body of permeable soil or rock that contains or transmits groundwater. Typically, aquifers fill or recharge from rain or snowmelt when the water flows downward until it reaches less permeable rock.

In times of drought or water scarcity, little water is available naturally to recharge existing groundwater supplies, which can become depleted by overuse. Groundwater withdrawals have tripled in the past 50 years, according to 2012 United Nations estimates cited in The World's Water: Volume 8. Areas with the highest groundwater withdrawals include parts of China, India, and the United States. Roughly 67% of all water withdrawn is destined for agricultural use, 22% is allocated for domestic use, and 11% goes for industrial use.

In some areas, including Australia and California, groundwater or aquifer recharging is being explored to help bolster water supplies. The process involves the injection or infiltration of excess surface water into underground aquifers. Water may be treated before it is injected. The water can be stored underground until it is needed. Some watersheds are being restored with native plant species in wetland areas to support aquifers' natural recharge capabilities.

Surface water is often stored in dams, lakes, reservoirs, and tanks, but there are many challenges associated with it, including flooding, pollution by natural and manmade sources, and losses from evaporation or seepage.

Water Reuse and Zero-liquid Discharge Technology

Several interrelated strategies and approaches to water reuse can alleviate water scarcity for municipalities and industries. These include water recycling and reuse, and the use of zero-liquid discharge (ZLD) systems, which use, treat, and reuse water in a closed-loop system without release or discharge.

Recycled, or reclaimed, water can be used in a variety of applications across industries, both inside facilities and in the community. Typical uses for recycled water include surface irrigation for orchards and vineyards, golf courses, landscaped areas, and food crops. Other uses include the recharging of groundwater, preservation or augmentation of ecosystems such as wetlands or riparian habitat, and in industrial processes. Nonpotable water can be used for toilet flushing, irrigating landscaping, washing vehicles and streets, and other similar purposes.

With these systems, wastewater — once viewed as a useless, disposable commodity — becomes a valuable resource. Fluence has worldwide experience in the advanced treatment of wastewater, creating systems for water reuse across a range of industrial, agricultural, and municipal processes. Fluence's water treatment technologies are capable of producing pure and ultrapure water for reuse in various applications, including power generation, beverage bottling, food production, and agriculture irrigation.

The Ashalim power plant concentrates sunlight to produce high-temperature steam for turbine generation.

One important aspect of water reuse is that it preserves valuable sources of fresh water. One example is the new Ashalim solar thermal power plant in Israel, which relies on local fresh water for its cooling-tower make-up water. The government was looking for a system to reuse fresh water and minimize the discharge of brackish blowdown wastewater.

Fluence devised a solution that includes filtration, ultrafiltration, and reverse osmosis in modular containers. Recycling the cooling water before discharge into evaporation ponds reduces use of valuable fresh water by 50% and lowers discharge volumes.

The food and beverage industry also uses water reuse and zero-discharge technologies. In fact, such technologies can improve their overall cost of operations as well as make them resilient in periods of water scarcity. In a March 2012 interview with Food Manufacturing, Henry J. Charrabé, managing director and chief executive officer of Fluence, explained:

> Food plants require a large volume of water to process foods, clean plant equipment and remove waste products. The enormous amount of wastewater that must be treated is a burdensome cost for many food manufacturers. This is why water and wastewater treatment present both a challenge and an opportunity for food plant operators.

A PepsiCo Frito-Lay facility in Casa Grande, Arizona, is reportedly the first U.S. food processing plant able to produce drinking-quality process water for reuse. The snack food manufacturing plant, which processes potatoes and corn, has a 2,460-m^3/d process water recovery treatment system that has helped Frito-Lay reduce its annual water use by 378,541 m^3. It landfills less than 1% of its waste, making it a near-net-zero waste facility.

Water reuse — whether it is grey water or recycled water — can save fresh water for human consumption in times of water stress and water scarcity. In Australia, for example, grey water use would reportedly save more than 1 trillion liters of fresh drinking water annually. Although some consumers are skeptical about drinking recycled water, vocal advocates — including Microsoft founder turned philanthropist Bill Gates — continue to demonstrate there is nothing to fear from drinking properly treated water.

Desalination

An increasingly popular solution to fresh water scarcity is treating saline or brackish water sources through a process known as desalination. This process can treat seawater or groundwater containing salt concentrations that make the water unfit for human consumption. Fresh water, for example, is defined as water with less than 1,000 ppm of salt. Highly saline water contains between 10,000 ppm and 35,000 ppm of salt.

Many nations are increasing their investment in desalination to develop reliable water sources in the face of growing demand. These include the United Arab Emirates, nations with limited available water supplies such as Cyprus, and water-stressed areas of the U.S. There are an estimated 16,000 desalination plants in operation around the world, the largest of which are in Saudi Arabia, the UAE, and Israel.

In the UAE, for example, water demand is expected to double between 2011 and 2020. Most of that demand is being filled through desalination, with roughly US$3.27 billion spent annually for desalination. Abu Dhabi reportedly produced 650 million GPD in 2011.

Unfortunately, desalination often has relied heavily on power-hungry, fixed facilities. Masdar estimated that seawater desalination requires about 10 times more energy than pumping well water does.

But there are solutions available to overcome these conventional obstacles. For example, some larger desalination facilities may include a cogeneration plant — a greener source of power for treatment.

Decentralized Treatment Solutions

Another way to save money is by bringing the treatment to where it's needed, eliminating the need for extensive delivery infrastructure. Fluence — which already designs, manufactures, and supplies state-of-the-art, full-scale desalination plants with capacities from 20,000 to 100,000 m³/d — is leading the way with affordable decentralized treatment solutions, like the modular, containerized water treatment plants in its NIROBOX line. Nirobox has options for both seawater and brackish water, as well as for wastewater treatment.

Nirobox is a plug-and-play, scalable solution that can be quickly deployed. One good example can be found at Reserva Conchal, a five-star resort in drought-affected Costa Rica. Water shortage and lack of water delivery infrastructure were posing a threat to tourism, necessitating a local, high-quality, dependable supply of potable water that wouldn't hurt the environment.

In just eight months from order to startup, Fluence provided three Nirobox containerized seawater desalination units with a total output of 1,500 m^3 of water a day.

Access to power can be an obstacle to decentralized water treatment. Conventional activated sludge treatment requires a substantial amount of energy to power aerators.

A revolutionary new technology from Fluence cuts energy use for aeration by 90%. Membrane aerated biofilm reactors (MABRs) not only save money, but their low energy needs make it possible to establish wastewater treatment in remote locations.

For example, an agricultural community of 1,000 homes in Israel's Jezreel Valley needed to update a pond system that was unable to lower nutrient levels to government standards. The solution had to be odorless and quiet, have low power consumption, and use the existing pond structure. By adding decentralized MABR units, the process was able to provide 125 m^3/d of water suitable for reuse in irrigation.

Water Management

The management of water resources using existing policies and regulations is a way to address many water-related challenges, including water reuse water rights, and others. It addresses the effects of natural events and human intervention — such as damming or dredging — on natural water resources, and also addresses the long-term, cumulative effects of water policy decisions on the economy, institutions, and environment. This may be through the development of policies regarding domestic water supplies, the pollution and overdrafting of groundwater supplies, wetlands restoration, and issues such as water imports and exports.

Although water management is commonly viewed as a task for national or regional governments, it is increasingly practiced at the state, provincial, or local level. Companies and industries are also adopting water management best practices to help them thrive and become better resource stewards.

One of the biggest obstacles limiting effective water management is politics and bureaucracy. A prime example can be seen in the Western U.S., where increased demand

and scarcity are making state and regional officials increasingly protective of their water rights.

Infrastructure Monitoring and Repairs

Another key in the water savings puzzle is the ongoing need worldwide for infrastructure monitoring and repair to prevent loss of water through delivery systems. These small amounts become increasingly larger over time. Monitoring aging infrastructure and creating new technologies — such as wireless smart valves and pipe defect and leak-detection sensing devices — are helping, but they must be used along with water policies such as routine reporting and repair plans.

How big is the problem? According to a 2013 report from the Center for Neighborhood Technology, in the U.S., an estimated 2.1 trillion gallons per year — about 16% of the water used in the nation daily — is lost through outdated and leaky infrastructure. In Europe, the estimated value of water lost through leaky infrastructure is roughly 80 billion euros per year, according to the Community Research and Development Information Service.

Contributing to the problem is inadequate funding for infrastructure repair or replacement. Even in the U.S., investments of more than $1 trillion are needed to repair and expand the nation's aging drinking water infrastructure, according to a 2013 American Water Works Association report. Estimates for repairing and upgrading wastewater treatment systems throughout the nation were similar.

The organization also noted that delaying investments on key infrastructure repairs dramatically increases the eventual costs. To address these issues in some areas, water utility privatization has been advocated. The World Bank, for example, estimated that public-private partnerships resulted in reducing water losses — from leaks, theft, and inaccurate measurement — by 15%.

Decentralized treatment can play an important role in reducing this type of water loss. By eliminating vulnerable infrastructure, it can lead to a reduction in so-called non-revenue water, which refers to water that's treated but lost through leaks or theft before reaching consumers.

Water Conservation

Water conservation is critical to stemming water scarcity. Although there are concerns about its effectiveness, it is needed to reduce demand. Typically, conservation efforts are publicized and encouraged in times of drought, but in reality, conservation is key to sustaining the supply-demand balance, especially in areas facing population growth.

Effective conservation efforts can be seen in areas such as Zaragoza, Spain, which instituted its Water Saving City project in 1997 with a goal of reducing domestic water use by 1 million m^3/y. The net effect has been a "water scarcity impact" of 1.176 m^3 of water per year, according Water 2030. This is a per-capita water use reduction of roughly 51 liters, or a change from 150 L/d in 1997 to 99 L/d in 2012, despite a 12% population increase.

Despite this and similar successes, conservation is frequently pummeled in the environmental media for being ineffective, especially in the absence of meaningful water management policy and low water prices. Kurt Schwabe, professor of environmental economics and policy at the University of California Riverside, was quoted in The Redlands Daily Facts as saying that critics say even the success of mandatory water restrictions is "a function of the good will of the public, also the probability of getting caught misbehaving."

The founder of the Environmentalist Foundation of India, Arun Krishnamurthy, observed in The Guardian:

> Most conservation efforts start with a bang, and fizzle out over the months or years due to a lack of support. This could be a lack of money, of public awareness, or even the in-depth knowledge needed to proceed further.

Ultimately, addressing water scarcity requires the combined efforts of consumers, water managers, researchers, and public officials.

WATER POLLUTION

Over two thirds of Earth's surface is covered by water; less than a third is taken up by land. As Earth's population continues to grow, people are putting ever-increasing pressure on the planet's water resources. In a sense, our oceans, rivers, and other inland waters are being "squeezed" by human activities—not so they take up less room, but so their quality is reduced. Poorer water quality means water pollution.

We know that pollution is a human problem because it is a relatively recent development in the planet's history: before the 19th century Industrial Revolution, people lived more in harmony with their immediate environment. As industrialization has spread around the globe, so the problem of pollution has spread with it. When Earth's population was much smaller, no one believed pollution would ever present a serious problem. It was once popularly believed that the oceans were far too big to pollute. Today, with around 7 billion people on the planet, it has become apparent that there are limits. Pollution is one of the signs that humans have exceeded those limits.

Water pollution can be defined in many ways. Usually, it means one or more substances have built up in water to such an extent that they cause problems for animals or people. Oceans, lakes, rivers, and other inland waters can naturally clean up a certain amount of pollution by dispersing it harmlessly. If you poured a cup of black ink into a river, the ink would quickly disappear into the river's much larger volume of clean water. The ink would still be there in the river, but in such a low concentration that you would not be able to see it. At such low levels, the chemicals in the ink probably would not present any real problem. However, if you poured gallons of ink into a river every few seconds through a pipe, the river would quickly turn black. The chemicals in the ink could very quickly have an effect on the quality of the water. This, in turn, could affect the health of all the plants, animals, and humans whose lives depend on the river.

Thus, water pollution is all about quantities: how much of a polluting substance is released and how big a volume of water it is released into. A small quantity of a toxic chemical may have little impact if it is spilled into the ocean from a ship. But the same amount of the same chemical can have a much bigger impact pumped into a lake or river, where there is less clean water to disperse it.

Pollution means adding substances to the environment that don't belong there—like the air pollution from this smokestack. Pollution is not always as obvious as this, however.

Water pollution almost always means that some damage has been done to an ocean, river, lake, or other water source. A 1969 United Nations report defined ocean pollution as:

> The introduction by man, directly or indirectly, of substances or energy into the marine environment (including estuaries) resulting in such deleterious effects as harm to living resources, hazards to human health, hindrance to marine activities, including fishing, impairment of quality for use of sea water and reduction of amenities.

Fortunately, Earth is forgiving and damage from water pollution is often reversible.

Main Types of Water Pollution

When we think of Earth's water resources, we think of huge oceans, lakes, and rivers. Water resources like these are called surface waters. The most obvious type of water pollution affects surface waters. For example, a spill from an oil tanker creates an oil slick that can affect a vast area of the ocean.

Not all of Earth's water sits on its surface, however. A great deal of water is held in underground rock structures known as aquifers, which we cannot see and seldom think about. Water stored underground in aquifers is known as groundwater. Aquifers feed our rivers and supply much of our drinking water. They too can become polluted, for example, when weed killers used in people's gardens drain into the ground. Groundwater pollution is much less obvious than surface-water pollution, but is no less of a problem. In 1996, a study in Iowa in the United States found that over half the state's groundwater wells were contaminated with weed killers. You might think things would have improved since then, but, two decades on, all that's really changed is the name of the chemicals we're using. Today, numerous scientific studies are still finding weed killers in groundwater in worrying quantities: A 2012 study discovered glyphosate in 41 percent of 140 groundwater samples from Catalonia, Spain; scientific opinion differs on whether this is safe or not.

Surface waters and groundwater are the two types of water resources that pollution affects. There are also two different ways in which pollution can occur. If pollution comes from a single location, such as a discharge pipe attached to a factory, it is known as point-source pollution. Other examples of point source pollution include an oil spill from a tanker, a discharge from a smoke stack (factory chimney), or someone pouring oil from their car down a drain. A great deal of water pollution happens not from one single source but from many different scattered sources. This is called nonpoint-source pollution.

Point-source pollution comes from a single, well-defined place such as this pipe. Below: Nonpoint-source pollution comes from many sources. All the industrial plants alongside a river and the ships that service them may be polluting the river collectively.

When point-source pollution enters the environment, the place most affected is usually the area immediately around the source. For example, when a tanker accident occurs, the oil slick is concentrated around the tanker itself and, in the right ocean conditions, the pollution disperses the further away from the tanker you go. This is less likely to happen with nonpoint source pollution which, by definition, enters the environment from many different places at once.

Sometimes pollution that enters the environment in one place has an effect hundreds or even thousands of miles away. This is known as transboundary pollution. One example is the way radioactive waste travels through the oceans from nuclear reprocessing plants in England and France to nearby countries such as Ireland and Norway.

When is Water Polluted?

Some forms of water pollution are very obvious: Everyone has seen TV news footage of oil slicks filmed from helicopters flying overhead. Water pollution is usually less obvious and much harder to detect than this. But how can we measure water pollution when we cannot see it? How do we even know it's there?

There are two main ways of measuring the quality of water. One is to take samples of the water and measure the concentrations of different chemicals that it contains. If the chemicals are dangerous or the concentrations are too great, we can regard the water as polluted. Measurements like this are known as chemical indicators of water quality. Another way to measure water quality involves examining the fish, insects, and other invertebrates that the water will support. If many different types of creatures can live in a river, the quality is likely to be very good; if the river supports no fish life at all, the quality is obviously much poorer. Measurements like this are called biological indicators of water quality.

Causes of Water Pollution

Most water pollution doesn't begin in the water itself. Take the oceans: around 80 percent of ocean pollution enters our seas from the land. Virtually any human activity can have an effect on the quality of our water environment. When farmers fertilize the fields, the chemicals they use are gradually washed by rain into the groundwater or surface waters nearby. Sometimes the causes of water pollution are quite surprising. Chemicals released by smokestacks (chimneys) can enter the atmosphere and then fall back to earth as rain, entering seas, rivers, and lakes and causing water pollution. That's called atmospheric deposition. Water pollution has many different causes and this is one of the reasons why it is such a difficult problem to solve.

Sewage

With billions of people on the planet, disposing of sewage waste is a major problem. According to 2015 figures from the World Health Organization, some 2.1 billion people (28 percent of the world's population) don't have access to safe drinking water, while

2.3 billion (30 percent of the world's population) don't have proper sanitation (hygienic toilet facilities); although there have been great improvements in securing access to clean water, relatively little progress has been made on improving global sanitation in the last decade. Sewage disposal affects people's immediate environments and leads to water-related illnesses such as diarrhea that kills 525,000 children under five each year. In developed countries, most people have flush toilets that take sewage waste quickly and hygienically away from their homes.

Yet the problem of sewage disposal does not end there. When you flush the toilet, the waste has to go somewhere and, even after it leaves the sewage treatment works, there is still waste to dispose of. Sometimes sewage waste is pumped untreated into the sea. Until the early 1990s, around 5 million tons of sewage was dumped by barge from New York City each year. According to 2002 figures from the UK government's Department for the Environment, Food, and Rural Affairs (DEFRA), the sewers of Britain collect around 11 billion liters of waste water every day; there are still 31,000 sewage overflow pipes through which, in certain circumstances, raw sewage is pumped untreated into the sea. The New River that crosses the border from Mexico into California once carried with it 20–25 million gallons (76–95 million liters) of raw sewage each day; a new waste water plant on the US-Mexico border, completed in 2007, substantially solved that problem. Unfortunately, even in some of the richest nations, the practice of dumping sewage into the sea continues. In early 2012, it was reported that the tiny island of Guernsey (between Britain and France) has decided to continue dumping 16,000 tons of raw sewage into the sea each day.

Nutrients

During crop-spraying, some chemicals will drain into the soil. Eventually, they seep into rivers and other watercourses.

Suitably treated and used in moderate quantities, sewage can be a fertilizer: It returns important nutrients to the environment, such as nitrogen and phosphorus, which plants and animals need for growth. The trouble is, sewage is often released in much greater quantities than the natural environment can cope with. Chemical fertilizers used by farmers also add nutrients to the soil, which drain into rivers and seas and add to the fertilizing effect of the sewage. Together, sewage and fertilizers can cause a massive increase in the growth of algae or plankton that overwhelms huge areas of oceans, lakes, or rivers. This is known as a harmful algal bloom (also known as an HAB or red tide, because it can turn the water red). It is harmful because it removes oxygen from the water that kills other forms of life, leading to what is known as a dead zone. The Gulf of Mexico has one of the world's most spectacular dead zones. Each summer, according to studies by the NOAA, it grows to an area of around 5500–6000 square miles (14,000–15,500 square kilometers), which is about the same size as the state of Connecticut.

Waste Water

A few statistics illustrate the scale of the problem that waste water (chemicals washed down drains and discharged from factories) can cause. Around half of all ocean pollution is caused by sewage and waste water. Each year, the world generates perhaps 5–10 billion tons of industrial waste, much of which is pumped untreated into rivers, oceans, and other waterways. In the United States alone, around 400,000 factories take clean water from rivers, and many pump polluted waters back in their place. However, there have been major improvements in waste water treatment recently. Since 1970, in the United States, the Environmental Protection Agency (EPA) has invested about $70 billion in improving water treatment plants that, as of 2015, serve around 88 percent of the US population (compared to just 69 percent in 1972). However, another $271 billion is still needed to update and upgrade the system.

Factories are point sources of water pollution, but quite a lot of water is polluted by ordinary people from nonpoint sources; this is how ordinary water becomes waste water in the first place. Virtually everyone pours chemicals of one sort or another down their drains or toilets. Even detergents used in washing machines and dishwashers eventually end up in our rivers and oceans. So do the pesticides we use on our gardens. A lot of toxic pollution also enters waste water from highway runoff. Highways are typically covered with a cocktail of toxic chemicals—everything from spilled fuel and brake fluids to bits of worn tires (themselves made from chemical additives) and exhaust emissions. When it rains, these chemicals wash into drains and rivers. It is not unusual for heavy summer rainstorms to wash toxic chemicals into rivers in such concentrations that they kill large numbers of fish overnight. It has been estimated that, in one year, the highway runoff from a single large city leaks as much oil into our water environment as a typical tanker spill. Some highway runoff runs away into drains; others can pollute groundwater or accumulate in the land next to a road, making it increasingly toxic as the years go by.

Chemical Waste

Detergents are relatively mild substances. At the opposite end of the spectrum are highly toxic chemicals such as polychlorinated biphenyls (PCBs). They were once widely used to manufacture electronic circuit boards, but their harmful effects have now been recognized and their use is highly restricted in many countries. Nevertheless, an estimated half million tons of PCBs were discharged into the environment during the 20th century. In a classic example of transboundary pollution, traces of PCBs have even been found in birds and fish in the Arctic. They were carried there through the oceans, thousands of miles from where they originally entered the environment. Although PCBs are widely banned, their effects will be felt for many decades because they last a long time in the environment without breaking down.

Another kind of toxic pollution comes from heavy metals, such as lead, cadmium, and mercury. Lead was once commonly used in gasoline (petrol), though its use is now restricted in some countries. Mercury and cadmium are still used in batteries (though some brands now use other metals instead). Until recently, a highly toxic chemical called tributyltin (TBT) was used in paints to protect boats from the ravaging effects of the oceans. Ironically, however, TBT was gradually recognized as a pollutant: boats painted with it were doing as much damage to the oceans as the oceans were doing to the boats.

The best known example of heavy metal pollution in the oceans took place in 1938 when a Japanese factory discharged a significant amount of mercury metal into Minamata Bay, contaminating the fish stocks there. It took a decade for the problem to come to light. By that time, many local people had eaten the fish and around 2000 were poisoned. Hundreds of people were left dead or disabled.

Radioactive Waste

People view radioactive waste with great alarm—and for good reason. At high enough concentrations it can kill; in lower concentrations it can cause cancers and other illnesses. The biggest sources of radioactive pollution in Europe are two factories that reprocess waste fuel from nuclear power plants: Sellafield on the north-west coast of Britain and Cap La Hague on the north coast of France. Both discharge radioactive waste water into the sea, which ocean currents then carry around the world. Countries such as Norway, which lie downstream from Britain, receive significant doses of radioactive pollution from Sellafield. The Norwegian government has repeatedly complained that Sellafield has increased radiation levels along its coast by 6–10 times. Both the Irish and Norwegian governments continue to press for the plant's closure.

Oil Pollution

When we think of ocean pollution, huge black oil slicks often spring to mind, yet these spectacular accidents represent only a tiny fraction of all the pollution entering our oceans. Even considering oil by itself, tanker spills are not as significant as they might seem: only 12 percent of the oil that enters the oceans comes from tanker accidents; over 70 percent of oil pollution at sea comes from routine shipping and from the oil people pour down drains on land. However, what makes tanker spills so destructive is the sheer quantity of oil they release at once — in other words, the concentration of oil they produce in one very localized part of the marine environment. The biggest oil spill in recent years (and the biggest ever spill in US waters) occurred when the tanker Exxon Valdez broke up in Prince William Sound in Alaska in 1989. Around 12 million gallons (44 million liters) of oil were released into the pristine wilderness—enough to fill your living room 800 times over! Estimates of the marine animals killed in the spill vary from approximately 1000 sea otters and 34,000 birds to as many as 2800 sea otters and 250,000 sea birds. Several billion salmon and herring eggs are also believed to have been destroyed.

Oil-tanker spills are the most spectacular forms of pollution and the ones that catch public attention, but only a fraction of all water pollution happens this way.

Plastics

If you've ever taken part in a community beach clean, you'll know that plastic is far and away the most common substance that washes up with the waves. There are three reasons for this: Plastic is one of the most common materials, used for making virtually every kind of manufactured object from clothing to automobile parts; plastic is light and floats easily so it can travel enormous distances across the oceans; most plastics are not biodegradable (they do not break down naturally in the environment), which means that things like plastic bottle tops can survive in the marine environment for a long time. (A plastic bottle can survive an estimated 450 years in the ocean and plastic fishing line can last up to 600 years).

While plastics are not toxic in quite the same way as poisonous chemicals, they nevertheless present a major hazard to seabirds, fish, and other marine creatures. For example, plastic fishing lines and other debris can strangle or choke fish. (This is sometimes called ghost fishing.) About half of all the world's seabird species are known to have eaten plastic residues. In one study of 450 shearwaters in the North Pacific, over 80 percent of the birds were found to contain plastic residues in their stomachs. In the early 1990s, marine scientist Tim Benton collected debris from a 2km (1.5 mile) length of beach in the remote Pitcairn islands in the South Pacific. His study recorded approximately a thousand pieces of garbage including 268 pieces of plastic, 71 plastic bottles, and two dolls heads.

Today, much media attention focuses on the Great Pacific Garbage Patch, a floating, oceanic graveyard of plastic junk roughly three times the size of France, discovered by sailor Charles J. Moore in 1997. But, as you'll know well enough if you've ever taken part in a community beach clean, persistent plastic litters every ocean on the planet: Some 8 million tons of new plastic are dumped in the sea every single year.

Alien Species

Most people's idea of water pollution involves things like sewage, toxic metals, or oil slicks, but pollution can be biological as well as chemical. In some parts of the world, alien species are a major problem. Alien species (sometimes known as invasive species) are animals or plants from one region that have been introduced into a different ecosystem where they do not belong. Outside their normal environment, they have no natural predators, so they rapidly run wild, crowding out the usual animals or plants that thrive there. Common examples of alien species include zebra mussels in the Great Lakes of the USA, which were carried there from Europe by ballast water (waste water flushed from ships). The Mediterranean Sea has been invaded by a kind of alien algae called Caulerpa taxifolia. In the Black Sea, an alien jellyfish called Mnemiopsis leidyi reduced fish stocks by 90 percent after arriving in ballast water. In San Francisco Bay, Asian clams called Potamocorbula amurensis, also introduced by ballast water, have dramatically altered the ecosystem. In 1999, Cornell University's David Pimentel estimated that alien invaders like this cost the US economy $123 billion a year; in 2014, the European Commission put the cost to Europe at €12 billion a year and "growing all the time.

Invasive species: Water hyacinth crowding out a waterway around an old fence post.

Other Forms of Pollution

These are the most common forms of pollution—but by no means the only ones. Heat or thermal pollution from factories and power plants also causes problems in rivers. By raising the temperature, it reduces the amount of oxygen dissolved in the water, thus also reducing the level of aquatic life that the river can support.

Another type of pollution involves the disruption of sediments (fine-grained powders) that flow from rivers into the sea. Dams built for hydroelectric power or water reservoirs can reduce the sediment flow. This reduces the formation of beaches, increases coastal erosion (the natural destruction of cliffs by the sea), and reduces the flow of nutrients from rivers into seas (potentially reducing coastal fish stocks). Increased sediments can also present a problem. During construction work, soil, rock, and other fine powders sometimes enters nearby rivers in large quantities, causing it to become turbid (muddy or silted). The extra sediment can block the gills of fish, effectively suffocating them. Construction firms often now take precautions to prevent this kind of pollution from happening.

Effects of Water Pollution

Some people believe pollution is an inescapable result of human activity: they argue that if we want to have factories, cities, ships, cars, oil, and coastal resorts, some degree of pollution is almost certain to result. In other words, pollution is a necessary evil that people must put up with if they want to make progress. Fortunately, not everyone agrees with this view. One reason people have woken up to the problem of pollution is that it brings costs of its own that undermine any economic benefits that come about by polluting.

Take oil spills, for example. They can happen if tankers are too poorly built to survive accidents at sea. But the economic benefit of compromising on tanker quality brings an economic cost when an oil spill occurs. The oil can wash up on nearby beaches, devastate the ecosystem, and severely affect tourism. The main problem is that the people who bear the cost of the spill (typically a small coastal community) are not the people who caused the problem in the first place (the people who operate the tanker). Yet, arguably, everyone who puts gasoline (petrol) into their car—or uses almost any kind of petroleum-fueled transport—contributes to the problem in some way. So oil spills are a problem for everyone, not just people who live by the coast and tanker operates.

Sewage is another good example of how pollution can affect us all. Sewage discharged into coastal waters can wash up on beaches and cause a health hazard. People who bathe or surf in the water can fall ill if they swallow polluted water—yet sewage can have other harmful effects too: it can poison shellfish (such as cockles and mussels) that grow near the shore. People who eat poisoned shellfish risk suffering from an acute—and sometimes fatal—illness called paralytic shellfish poisoning. Shellfish is no longer caught along many shores because it is simply too polluted with sewage or toxic chemical wastes that have discharged from the land nearby.

Pollution matters because it harms the environment on which people depend. The environment is not something distant and separate from our lives. It's not a pretty shoreline hundreds of miles from our homes or a wilderness landscape that we see only on TV. The environment is everything that surrounds us that gives us life and health. Destroying the environment ultimately reduces the quality of our own lives—and that, most selfishly, is why pollution should matter to all of us.

Preventing Water Pollution

There is no easy way to solve water pollution; if there were, it wouldn't be so much of a problem. Broadly speaking, there are three different things that can help to tackle the problem—education, laws, and economics—and they work together as a team.

Education

Making people aware of the problem is the first step to solving it. In the early 1990s, when surfers in Britain grew tired of catching illnesses from water polluted with sewage, they formed a group called Surfers Against Sewage to force governments and water companies to clean up their act. People who've grown tired of walking the world's polluted beaches often band together to organize community beach-cleaning sessions. Anglers who no longer catch so many fish have campaigned for tougher penalties against factories that pour pollution into our rivers. Greater public awareness can make a positive difference.

Laws

One of the biggest problems with water pollution is its transboundary nature. Many rivers cross countries, while seas span whole continents. Pollution discharged by factories in one country with poor environmental standards can cause problems in neighboring nations, even when they have tougher laws and higher standards. Environmental laws can make it tougher for people to pollute, but to be really effective they have to operate across national and international borders. This is why we have international laws governing the oceans, such as the 1982 UN Convention on the Law of the Sea (signed by over 120 nations), the 1972 London (Dumping) Convention, the 1978 MARPOL International Convention for the Prevention of Pollution from Ships, and the 1998 OSPAR Convention for the Protection of the Marine Environment of the North East Atlantic. The European Union has water-protection laws (known as directives) that apply to all of its member states. They include the 1976 Bathing Water Directive, which seeks to ensure the quality of the waters that people use for recreation. Most countries also have their own water pollution laws. In the United States, for example, there is the 1972 Clean Water Act and the 1974 Safe Drinking Water Act.

Economics

Most environmental experts agree that the best way to tackle pollution is through something called the polluter pays principle. This means that whoever causes pollution should have to pay to clean it up, one way or another. Polluter pays can operate in all kinds of ways. It could mean that tanker owners should have to take out insurance that covers the cost of oil spill cleanups, for example. It could also mean that shoppers should have to pay for their plastic grocery bags, as is now common in Ireland, to encourage recycling and minimize waste. Or it could mean that factories that use rivers must have their water inlet pipes downstream of their effluent outflow pipes, so if they cause pollution they themselves are the first people to suffer. Ultimately, the polluter pays principle is designed to deter people from polluting by making it less expensive for them to behave in an environmentally responsible way.

OVERDRAFTING

Overdrafting is the process of extracting groundwater beyond the equilibrium yield of the aquifer.

There are two sets of yields, safe yield and sustainable yield. Safe yield is the amount of water that can be taken out of the ground without there being any undesirable results. Sustainable yield is extraction that takes into account both recharge rate and surface water impacts.

There are two types of aquifers: confined and unconfined. In confined aquifers, there is an overbearing layer called aquitard, which contains impermeable materials through which groundwater cannot be extracted. In unconfined aquifers, there is no aquitard, and groundwater can be freely extracted from the surface. Extracting groundwater from unconfined aquifers is like borrowing the water, it has to be recharged at a proper amount. If recharge is not done in proper amounts there can be many impacts. Recharge may happen through artificial recharge and natural recharge.

Natural process of recharge is done through percolation of surface water. Artificial process of recharging the aquifer is through means of pumping reclaimed water from wastewater management projects directly into the aquifer. An example is the Orange County Water District in the State of California. This organization take waste water, treats it to a proper level, and then systematically pumps it back into the aquifers for artificial recharge.

When groundwater is extracted the water is primarily pulled from the aquifer which creates a cone depression around the well. When drafting of water continues the cone of depression increases in width. The increase in width leads to the negative impacts caused by overdrafting, such as drop of the water table, land subsidence, and loss of surface water reaching the streams. In extreme cases the supply of water to naturally recharge the aquifers is pulled directly from streams and rivers, leading to depletion of water levels in streams and rivers. The depletion of water in rivers and streams has an effect on wildlife, as well as humans who might be using the water for other purposes.

Since every groundwater basin recharges at a different rate depending upon precipitation, vegetative cover and soil conservation practises, the quantity of groundwater that can be safely pumped varies greatly among regions of the world and even within provinces. Some aquifers require a very long time to recharge and thus the process of overdrafting can have consequences of effectively drying up certain sub-surface water supplies. Subsidence occurs when excessive groundwater is extracted from rocks that support more weight when saturated. This can lead to a capacity reduction in the aquifer.

Groundwater is the fresh water that can be found underground; it is also one of the largest sources. Groundwater depletion can be comparable to “money in a bank”, The primary cause of groundwater depletion is pumping or the excessive pulling up of groundwater from underground aquifers.

Around the World

Ranking of countries that use groundwater for irrigation.	
Country	Million hectares (1×10^6 ha (2.5×10^6 acres)) irrigated with groundwater
India	26.5

USA	10.8
China	8.8
Pakistan	4.9
Iran	3.6
Bangladesh	2.6
Mexico	1.7
Saudi Arabia	1.5
Italy	0.9
Turkey	0.7
Syria	0.6
Brazil	0.5

The ranking is based on the amount of groundwater each country uses for agriculture. This issue is becoming quite large in the United States (most notably California), but it is also worth noting that it has been a problem in other parts of the world, as was documented in Punjab, India in 1987.

Accelerated Decline in Subterranean Reservoirs

According to a 2013 report by research hydrologist, Leonard F. Konikow, at the United States Geological Survey (USGS), the depletion of the Ogallala Aquifer between 2001–2008, inclusive, is about 32 percent of the cumulative depletion during the entire 20th century (Konikow 2013:22)." In the United States, the biggest users of water from aquifers include agricultural irrigation and oil and coal extraction. "Cumulative total groundwater depletion in the United States accelerated in the late 1940s and continued at an almost steady linear rate through the end of the century. In addition to widely recognized environmental consequences, groundwater depletion also adversely impacts the long-term sustainability of groundwater supplies to help meet the Nation's water needs."

According to another USGS study of withdrawals from 66 major US aquifers, the three greatest uses of water extracted from aquifers were agriculture (irrigation) (68%), public water supply (19%), and self-supplied industrial (4%). The remaining about 8% of groundwater withdrawals were for "self-supplied domestic, aquaculture, livestock, mining, and thermoelectric power uses."

Impacts on the Environment

The environmental impact of overdrafting includes:

- Land subsidence - The collapse of land from lack of support (from the water that is being depleted). The first recorded case of land subsidence was in the 1940s. Land subsidence can be as little as a local land collapsing or as large as an entire region's land being lowered. The subsidence can lead to infrastructural and ecosystem damage.
- Lowering of the water table, which makes water harder to reach streams and rivers.
- Reduction of water volume in streams and lakes because their supply of water is being diminished by surface water recharging the aquifers.
- Animals that depend on streams and lakes for food and water and habitat will be affected.
- Deteriorating water quality.
- The cost of water to the consumer rises. This is due to the water table lowering so more energy is needed to pump further down, and to cover that pumping companies need more profit.
- Crop production decrease from lack of water(60% of US irrigation relies on groundwater so this is a large loss).
- Groundwater depletion also throws off the water cycle.

Effects on Climate

Aquifer drawdown or overdrafting and the pumping of fossil water may be a contributing factor to sea-level rise. By increasing the amount of moisture available to fall as precipitation, severe weather events are more likely to occur. To some extent moisture in the atmosphere accelerates the probability of a global warming event. The correlation coefficient is not yet scientifically determined.

Socio-economic Effects

Scores of countries are overpumping aquifers as they struggle to satisfy their growing water needs, including each of the big three grain producers— China, India, and the United States. These three, along with a number of other countries where water tables are falling, are home to more than half the world's people.

Water is intrinsic to biological and economic growth, and overdraft limits its available supply. According to Liebig's law of the minimum, growth is therefore impeded. Deeper wells must be drilled as the water table drops, which can become expensive. In addition, the energy needed to extract a given volume of water increases with the amount the aquifer has been depleted. The deeper the water is extracted from the worse the quality of the water becomes, which increases the cost of filtration. Saltwater intrusion is another consequence of overdrafting, leading to a reduction in water quality.

Possible Solutions

- Artificial recharge - Since recharge is the natural replenishment of water, *artificial* recharge is the artificial or man made replenishment of groundwater. This is the more aesthetically pleasing option though there is a limited amount of water that can be used for replenishing which is a problem.
- Decreased water use - This can be used in combination with the solution above. Consumptive use refers to the water that is naturally taken from the system (e.g., transpiration). So decreasing that use is what needs to happen in areas where recharge alone will not work. It requires such things as switching to less water-intensive crops.

AGRICULTURE AND WATER SCARCITY

Agriculture is both a victim and a cause of water scarcity. Water of appropriate quality and quantity is essential for the production of crops, livestock, and fisheries, as well as for the processing and preparation of these foods and products. Water is the lifeblood of ecosystems, including forests, lakes, and wetlands, on which the food and nutritional security of present and future generations depends. At the same time, agriculture is the largest water user globally, and a major source of water pollution. Unsustainable agricultural water use practices threatens the sustainability of livelihoods dependent on water and agriculture.

Additionally, climate change will have significant impacts on agriculture by increasing water demand, limiting crop productivity, and reducing water availability in areas where irrigation is most needed or has a comparative advantage. A growing number of regions will face increasing water scarcity. Climate change will bring greater variation in weather events, more frequent weather extremes, and new challenges requiring the sector to take mitigation and adaptation actions.

What can agriculture do to address water scarcity in the context of climate change, while ensuring food and nutrition security? What responses can the agriculture and food sectors offer to alleviate the impacts – and reduce the risks – of water scarcity? These questions were at the center of the Agriculture and Food Security Action Day during the COP22 summit in Marrakesh in November.

In a bid to tackle the impact of global water scarcity, FAO launched the Global Framework for Action to Cope with Water Scarcity in Agriculture in the Context of Climate Change. This initiative is based on the premise that a sustainable pathway to food security, in the context of water scarcity, lies in maximizing benefits that cut across multiple dimensions of the food–water–climate nexus. Appropriate responses to water scarcity will be found not only in the water domain, but also in all the agriculture and food sectors– crop production, livestock, fisheries, aquaculture, and forestry. Most water is withdrawn at the production stage, but water scarcity can also be addressed along food value chains and by consumers. Agriculture holds the key to coping with water scarcity as it is responsible for 70% of all freshwater withdrawals.

Farmers worldwide have been testing solutions to water scarcity, but they need to be supported with appropriate policies, the right mix of public and private investments, and access to knowledge and resources for producing more and better with less water. Various adaptation measures that deal with climate variability and build on improved land and water management practices have the potential to create resilience to climate change and address water scarcity. The sustainable intensification of food production, with more efficient water management systems adapted to climate variability and local circumstances, can help increase water productivity and raise on-farm incomes. Countries in water-scarce regions will increasingly need to devise food security strategies that explicitly consider structural food supply deficits, and trade arrangements that will provide protection from food price volatility.

FAO's Global Framework for Action will assist vulnerable countries to strengthen their capacities to adapt agriculture to the impacts of climate change and water scarcity, and implement their nationally determined contributions (NDCs) related to agriculture and water. The FAO Water Platform will support the implementation of the initiative.

Building on FAO's previous initiatives on water scarcity, the Global Framework is a call for international partners to contribute to two working groups on knowledge and innovation, and policies and investments. The initiative will focus on a range of important thematic areas to address issues of water scarcity in agriculture, including:

- Sustainable improvements of agricultural water productivity, cutting across all agricultural subsectors, from crop to livestock production, aquaculture and agroforestry, based on introduction of best practices in soil and water management, complementary afforestation, and sustainable grazing management.
- Modernization and development of multipurpose and climate proofing irrigation infrastructure are considered also as important action areas to improve the efficiency of water use in agriculture while adapting to climate change impacts.
- Integrated landscape approaches may play critical roles in regulating rainfall and other climate patterns and influencing water yields, sediment levels and water quality, and increasing water availability for agricultural use.

Other suggested areas of collaboration include:

- The reuse of treated wastewater for food production while ensuring food safety and prevention and controlling of water pollution from agricultural activities.
- More efficient use of water along food value chain has a good potential for water savings, reduction of water pollution and producing better quality food products. As about one-third of all food produced for human consumption in the world is lost or wasted each year, actions to prevent, reduce, reuse and recycle food losses and waste are part of the solution to using water resources more effectively and reducing carbon and blue-water footprints.
- Diversification of production systems and income opportunities for smallholders may provide options for less water-intensive production and more climate-resilient livelihoods.
- In the situation of growing water scarcity, in many countries food security will increasingly depend on food trade. A collective effort at an international level is therefore required to address the trade–food–water nexus and applications of the virtual water concept.

OVERPOPULATION

Water is one of the most integral and important aspects of daily life for every human being, for example, food, clothing, and almost everything else humans interact with involves water. Therefore, water and water security is going to be a crucial focus for governments in the next few decades, especially since the population is expected to reach approximately 9.7 billion by the year 2050, and 11.2 billion by 2100. Similar to oil and other fossil fuels, water is a finite resource, and the knowledge for world leaders to be able to manage a limited resource with a growing population will be critical for to have in order to maintain or grow their nations' prosperity. On the other hand, if current water resources are not properly regulated, an eventual increase in world population will become problematic for many regions and countries. Overpopulation will strain current water resources to their limits, cause an increase in water pollution, and lead to an increase in civil and international conflicts over existing water supplies.

One of the consequences of overpopulation is the pressure that is put on available water resources in order to serve a growing population. Approximately fifty percent of the worlds' population will be living in regions around the globe that are considered "water stressed", a term defined as when the demand for water exceeds the amount that is available, either due to lack of it, or poor quality, by the year 2030 (compared to fifteen percent currently). Since 1990, the global population increased by an average of eighty million people, which heightens the world demand for freshwater by about sixty four billion cubic meters of water per year. This increasingly high demand for water will also affect food production in water stressed areas such as the Middle East, India, China, and the southwestern United States. Water intensive crops in California, such as almonds, use approximately eight percent of all available freshwater, and one ton of grain requires one thousand tons of water. Worldwide grain and staple crop production uses between seventy five to ninety percent of accessible freshwater. Places in the Middle East, India, and China are going to experience at least and additional fifty million people of each of their populations without adequate food by 2050. The connections between a growing population that needs a higher demand for drinking water and water for agriculture shows that the shortages of water that are expected to affect many regions of the world will have severe consequences on the lives of millions of people, and that world leaders will need to find solutions in order to conserve and protect water resources for their countries, or find alternative methods to find new sources of water, such as desalination.

Growths in regional and global population will also lead to increased cases of water pollution. As of 2013, there are an estimated seven hundred and eighty million people who don't have access to safe drinking water, while about two billion people don't have proper water sanitation. About half of these statistics are for people living in cites. From the projected population size of around ten billion by 2050, the number of people who will live in urban areas is expected to increase almost two and a half billion people by that same year, on top of almost four billion people currently, putting the global urban population at about sixty five percent in 2050. Urban areas have a high risk of water pollution. Runoff from streets can carry oils, heavy metals, and other containments, while sewage water can leak into ground water, bringing bacteria, nitrates, phosphorus and other chemicals. Waste dumping also can pollute existing sources of freshwater with hazardous materials and toxic chemicals. It is estimated that between forty to fifty percent of all available freshwater sources on earth are polluted. The combination of the expected substantial increase of people residing in urban areas and the preexisting dangers of water pollution in urban settings, will lead to a rise in the amount of water that is not potable due to pollution. It is imperative that infrastructure to limit freshwater pollution is invested in in the future, by both developed and underdeveloped nations.

Finally, the pressures that are put on water resources by overpopulation will lead to civil and international conflict over control of available quantities. Accounts of battles and fights over water resources dates back to 600 BC, when Assyrians would poison, divert, and destroy water supplies in order to put their enemies under siege. Since the year 2000, there have been at least over one hundred and ten major conflicts over water resources either between nations or within one. Middle Eastern countries, such as Iraq, Iran, Afghanistan, Yemen, and Syria, countries in Africa like Darfur, Sudan, and Somalia, and the South American countries of Peru and Brazil have all experienced armed struggles involving scarce water supplies. These armed conflicts are due to multiple countries relying on a single water source, such as the

Shatt al-Arab river between Iran and Iraq. A dispute over water withdrawal from the river was an important factor that caused the Iran-Iraq war in 1980. The United States has listed water scarcity, and other consequences of overpopulation and climate change, as a threat to national security, realizing the important social and political chaos that limited water access can cause. To help dissuade countries from engaging in armed conflict over water resources, government leaders need to recognize how water is a finite resource, and the consequences that can happen when a finite resource is abused. There has to be agreements and contracts between nations who are in water stressed regions and who share the same water source that outline distribution amounts according to population size and or agricultural needs.

As the global population is expected to keep growing in the coming decades, the negative impact that humans will have on earths finite resources, especially water, will become increasingly apparent as areas of the world will start to experience drastic shortages of water, leading to instability in food production, industry, social order, and political and military control. In order to limit the amount of chaos and conflict that will ensue over limited water resources, there needs to be compromise and cooperation between all countries, not just the nations that are water stressed, to provide water management techniques, newer and more efficient technology to conserve as much water as possible, and strict security and enforcement of all regulations to prevent groups and individuals using water to gain power.

EFFECTS OF WATER SCARCITY

Hunger

Water is incredibly required to grow crops and to care for livestock animals. It is estimated that the global use of water for irrigation and agriculture is about 70% and that only 10% is utilized for domestic purposes. As a result, water shortage means the practice of growing crops and farming is greatly impacted. For this reason, water scarcity commonly contributes to lower yields and death of animals particularly in the arid and semiarid regions and as such, it results in hunger, poverty, and thirst.

Poor Heath

In many developing nations, water scarcity forces people to drink water of low quality from flowing streams, majority of which are contaminated. Accordingly, they are infected with water-borne diseases such as cholera, typhoid, and dysentery that kill people. Water shortage may also mean sewage systems are stagnant which creates room for the build-up of bacteria and harmful insects that result in infections. Besides, sanitation might become chaotic when water is scarce especially in restaurants, clinics, and public places thus compromising the health of the general public.

Poverty

Access to quality water is fundamental to better living standard and economic growth. Schools, restaurants, hospitals, hotels and other businesses need to stay clean for operations to run effectively. Imagine a situation whereby a major school or hotel goes without water even for a day, the situation can be disastrous and leads to enormous economic losses. Restaurants and shopping malls have to be kept clean to attract visitors. Manufacturing and industrial processes, mining activities, and commercial businesses all need large quantities of water to flourish. Without economic activities because of lack of water, then it means higher poverty levels and poor living standards.

Habitat Loss and Destruction to Ecosystems

When water is scarce, then it means the natural landscapes suffer the most as it contributes to desertification, lose of plants and death of wildlife and other animals. As a result, these ecological catastrophes create habitat loss that, in turn, leads to food shortages and poor quality of life. For instance, the Aral Sea in Central Asia that used to be the world's fourth largest freshwater lake has been reduced by more than a third in a period of only three decades. The water is now very salty, and the ecosystems within and around it have been extensively destroyed due to overuse of the water resource, mainly influenced by water scarcity in the region.

Disappearance of Wetlands

According to WWF, more than half of the planet's wetlands have lost since 1990 which is largely due to water scarcity. The wetlands have become dry to the point of losing its natural capability to hold water. Human activities are the main contributors because of water overuse, pollution, and interference with the underground aquifers.

SOLUTION FOR WATER SCARCITY

Over 2.7 billion people suffer from at least 1 month of water scarcity in a year.

These facts can be hard to grasp, especially since only 3% of the water in the world (which accounts for over 70% of the earth's surface) is freshwater deemed fit for human consumption.

However, when we factor in the fact that 2/3 of the freshwater in the world is locked up in frozen glaciers, we can understand how much of a problem water scarcity really is. Perspectively, while over a billion people spend days searching for clean and safe drinking water, some of us with access to plenty of water often take it for granted.

Major factors of water scarcity include pollution and overuse of water, conflicts, travel distance to water sources, drought, and governmental control. With water scarcity comes hunger, diseases and other health complications, poor sanitation, lack of education, and ultimately, poverty.

The good news is that the problem of water scarcity can be solved, but it requires a collective effort with each and every one of us playing our own part to solve this global crisis. Here are some proposed solutions to the water scarcity crisis.

Improved Infrastructure

Water infrastructure is a very important element of water management and control. It includes all the infrastructure used to build, pump, transport, divert, store, treat, and deliver safe drinking water, as well the tools and equipment used to build them. These structures include groundwater wells, dams, storage tanks, surface-water intakes, pipes, drinking-water facilities, and aqueducts. With poor water infrastructure inevitably comes water scarcity. This is why it is important for us to improve the quality of our water infrastructure be it natural or artificial.

Infrastructure also encompasses natural infrastructure which makes use of landscape management techniques such as conservation, restoration, and sustainable management. These techniques provide basic water services such as flood control, aquifer storage and recharge, and providing a clean and abundant supply of water. Improved infrastructure will not only reduce the amount of clean water being wasted but will also help reduce the number of persons constantly searching for clean water on a daily basis.

Groundwater Recharge/Aquifer Storage and Recovery

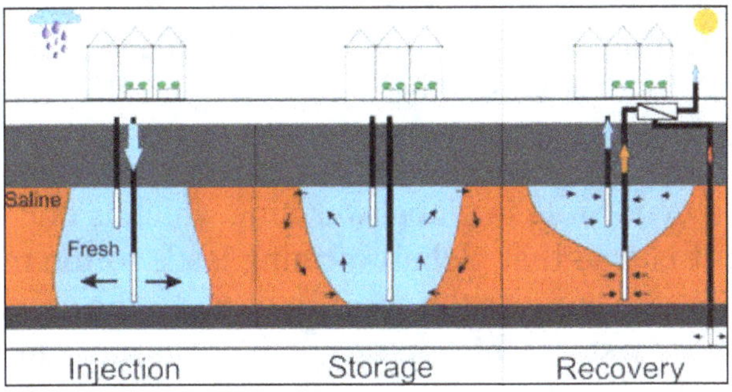

Another solution to the water scarcity crisis is to improve our attempts to recharge our groundwater. Artificial groundwater recharging has been in existence for decades and has proven to be one of the most successful methods of solving water scarcity in many areas. The process of recharge can either be a human-induced process or a natural process as part of the hydrological cycle. Groundwater recharge can be done directly via injection wells or spreading basins or as a result of human activities including waste disposal and irrigation. Reclaimed wastewater and human-induced recharge are gaining grounds in many regions, gradually making it a key component of the hydrological cycle.

Aquifer recovery and storage is another water scarcity solution. Here, recharge is done via surface infiltration into unconfined, shallow aquifers. In recent years, this process has evolved to accommodate direct injection into deeper aquifer systems, including those containing poor quality groundwater, making freshwater storage available in areas where none formerly existed. A major constraint to recharging by injection and infiltration is the clogging of pores from particulates, chemical precipitation, and biofilm formation. This can have adverse effects on the quality of groundwater such as the formation of dissolved sulfides and iron solubilization. That notwithstanding, if done and controlled properly, these two methods can greatly reduce the level of water scarcity in some areas stricken by this phenomenon.

Pollution and Contamination Control

Pollution is a major contributor to water scarcity in the world today. Water pollution, which is the dumping of harmful substances into clean water systems, is playing a major part in reducing the amount of water available for consumption. Individuals, as well as industries, keep polluting the water supply with chemicals and other forms of waste. With pollution comes a lot of adverse effects, especially health-related effects and eventually high mortality rates.

Pollution also plays a major role in global warming which is one of the major threats to our planet today. It is true that global warming leads to the melting of our frozen freshwater, but this water seeps into the ocean and causes the ocean levels to rise, leading to floods and other disasters which in turn reduce the amount of water in shortage (Kinhal). Air pollution leads to the contamination of rainwater which is a major source of water to many. It is up to us to reduce the insurmountable level of pollution plaguing our planet.

Water Conservation/Drought Mitigation

Water conservation is another solution to the current water crisis plaguing the globe. As was mentioned earlier, while many spend days searching for water, some of us have it in abundance but waste it, either intentionally or unintentionally. Here are some ways through which we can conserve water:

- Filling a sink basin with water when washing dishes instead of letting the water run.
- Using a bucket to catch the water while waiting for the shower to warm up.
- Avoid letting the water run when brushing teeth or applying soap on our skins when taking a bath.
- Installing water-saving fixtures and appliances.
- Replacing your high-flowing showerhead with low-flowing ones.
- Use a broom to clean sidewalks instead of water hoses.
- Eliminating leakages in bathtubs, faucets, and toilets.
- Flush your toilets only when necessary.

In the fight against water scarcity, every drop of water counts. As the poem by Julia Abigail Fletcher Carney goes,

"Little drops of water,
Little grains of sand,
Make the mighty ocean
And the pleasant land."

Drought mitigation is yet another technique to curtail the level of water scarcity in the world. Drought is a slow and gradual process which can be mitigated with proper preparation and with the right techniques. Drought mitigation techniques such as soil and water conservation and efficient irrigation can greatly reduce the effects of water scarcity.

Rainwater Harvesting/Water Catchment

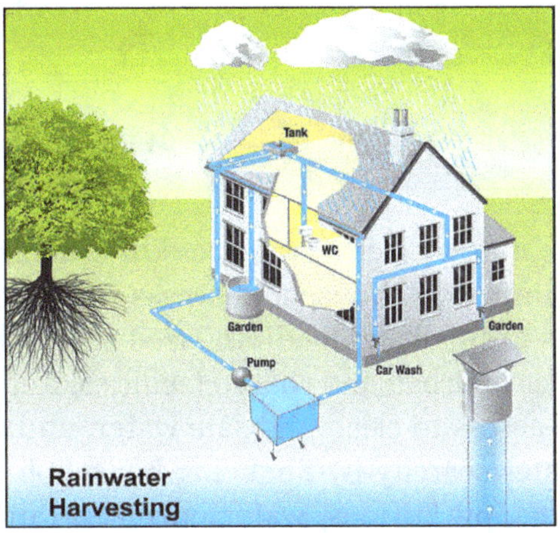

Rainwater harvesting is one of the main sources of water for a lot of people, especially in Africa. This is generally done via the use of a water catchment and collection basin. Water collected using a water catchment can be used to water our vegetable gardens, livestock, perform daily chores, cook, and even for drinking. Water collected directly from the atmosphere can be used for drinking, even without boiling. In many parts of Africa, water catchment and rainwater harvesting have proven to be a very important source of water to many. People build different catchment systems from roofs and gutters. Websites exist which help people with designs and different means of building great water catchment systems for harvesting rainwater.

A major advantage of this method is that it allows for the conservation of safe drinking water while the harvested rainwater is used to take care of daily chores. This method is also useful in areas where the natural landscape acts as a natural water catchment and favors the collection of rainwater. This makes it possible for rainwater to be harvested in large quantities. This harvested water can then be treated and supplied to the population such as in the case of the Greater Sydney area, Australia.

Water Credit/Water Equity

Mostly practiced in developing countries, this method of credit entails the provision of loans to households and small enterprises in an effort to increase access to a safe and improved water source as well as sanitation. Most water credit investments are financed by governments and other public-sector organizations, but due to the fact that their current investment levels aren't enough to reach everyone, private organizations and Micro Credit institutions have come in to complement the effort. Microcredits usually target the poor and less fortunate and help them to gain access to clean and improved water sources, safe water, and sanitation.

In the water sector, three major types of water credit schemes exist. These are water credit schemes aimed at improving household access to water supply, schemes aimed at improving water access to small and medium-sized enterprises, and schemes aimed at upgrading water services in urban and shared facilities. Regardless, the aim of water credit is to help improve access to clean and safe water, and to reduce the number of people suffering from water scarcity by financing household water and connections, toilets, latrines, bathrooms, pit latrines, water purifiers, rainwater harvesting tanks, and other systems.

Research and Technology

Recent breakthroughs in research and technology have come up with techniques that have made it possible for water considered unfit for consumption, to be made clean and safe for consumption. Some of the most frequently used techniques include Reverse Osmosis, Electrodialysis Reversal (EDR), Desalinization, Nanofiltration, and Solar and UV Filtration:

- Reverse osmosis uses semi-permeable membranes to remove molecules, ions and other larger particles from water, making it consumable. This method can be used to remove all types of suspended or dissolved species from water, including bacteria, thereby purifying water and making it potable. It has also been adapted and used in making household filters and other filtration devices.

- Unlike Reverse Osmosis which uses pressure, Electro-Dialysis Reversal (EDR) uses direct current. During the process, ions flow to electrodes of opposite charge through ion selective membranes. The polarity of electrodes used in EDR systems is periodically reversed. This method is used principally used when purifying waters with low total suspending solids (TSS) or high silica.
- Desalination is a process of water purification used to make salty water pure and clean by separating dissolved salts and other minerals. Desalination can also be used to purify brackish water, water from wells, and other forms of running water such as water from rivers and streams. This process which can be used to produce enough water capable of supporting large coastal populations is continually under evolving and under research to improve efficiency.
- Another method of water purification is Nanofiltration which seeks to remove two-valued ions, as well as other mono-valued ions such as heavy metals from water to make it pure and safe for consumption. This method is mainly used for softening water and removing specific heavy metals from water as well as in the reduction of salt contents in slightly brackish water. Because the method uses less fine membranes when compared to other techniques, its fouling rate and feed pressure are usually lower when compared to reverse osmosis systems.
- Solar and UV Filtration is a purification technique that uses a blend of solar energy and solar ultra-violet (UV) light to make water contaminated by biological organisms such as viruses, bacteria, worms, and protozoa, safe to drink.

Using these techniques can help us fight water scarcity and make sure that more people have access to clean and safe water.

Support Clean Water Initiatives

Another very important solution to water scarcity is supporting and donating to organization and initiatives that seek to provide clean water to those in need. Most of these organizations fund water projects using donations. Donations come from donors all over the world, they act as a middleman between those who do not have access to clean water and those who are willing to help. Some of these organizations include NGOs like Charity Water, Water.org, Lifewater International, WaterLex, Save The Water, World Water Council, and World Water Council.

Increase Awareness and Education

Another very important and simple solution to the problem of water scarcity is to increase awareness and education on water scarcity and its effects. There are a lot of people who are unaware of the reality of water scarcity as well as its negative effects and the number of people suffering from it. We must educate people on how to conserve water, on how to support water organizations and NGOs, and on the importance of water. Raising awareness and educating people on the reality of water scarcity, its effects and the importance of conserving water is an important step in the fight against water scarcity.

References

- What-is-water-scarcity: fluencecorp.com, Retrieved 1 January, 2020
- “Rising sea levels attributed to global groundwater extraction”. University of Utrecht. Retrieved February 8, 2011
- Waterpollution: explainthatstuff.com, Retrieved 2 February, 2020
- Lassiter, Allison (July 2015). Sustainable Water Challenges and Solutions from California. University of California. ISBN 9780520285354
- Guide-waterpollution-66615937-2400, media-uploads, full_content, styles, files, default, sites, assets.nrdc.org, Retrieved 3 March, 2020
- Media-uploads/guide-waterpollution-c09gbd-rm-ds-2400, public, full-content, styles, files, default, sites: assets.nrdc.org, Retrieved 4 April, 2020
- Agriculture-holds-key-tackling-water-scarcity, water: worldbank.org, Retrieved 5 May, 2020
- Causes-effects-and-solutions-to-water-scarcity, environment: eartheclipse.com, Retrieved 6 June, 2020

3

Water Conservation

Water conservation refers to the policies, strategies and activities performed for the sustainable management and protection of the water resources. Some of the methods used for water conservation are rainwater harvesting, desalination, drip irrigation, etc. The topics elaborated in this chapter will help in gaining a better perspective about the different methods and techniques used for water conservation.

Water conservation is the careful use and preservation of water supply. It includes both the quantity and quality of water utilized. Water is an essential asset for the nourishment of all life. The fundamental demand for all activities appropriates from local use to the agricultural industry.

With the regular expanding weight of the human population, there has been a serious tension on water resources. Negligence of customary water bodies like tanks and lakes, unpredictable and abuse of groundwater, and incorrect preservation of surface water systems have bothered the issue. Still further and is no doubt going to grow in the years to come.

There are various approaches to make your water to last nowadays. One simple yet often disregarded strategy to cut your water bill is to use your water twice. Unlike electricity, water can be reused over again and again. That's the idea of water conservation.

Key Facts about our Water

Water is definitely the most important natural resources that living things need. But at the same time, it has also been misused and wasted. To better grasp the total significance of water conservation, take a look at the few yet key facts about water:

- The average adult human body is composed of 50-65 percent of water: Averaging around 57-60 percent. With infants, they have a higher percentage. Often around 75-78% water, dropping to 65% by one year of age.

- The Earth has a limited amount of water: The water we have now is all we get, it is recycled over and over. The water cycle can help you understand this condition.
- Water is the basic demand for every food: It grows our fruits and vegetable and it is consumed by each livestock.
- Plant's life is dependent on water: Plants help the ecosystem and produce oxygen that is necessary to keep us healthy. Additionally, trees are generally used for housing, paper, and a lot more.

Water conservation has turned into an essential practice in every part of the world, even in regions where water appears to be enough. It is the most practical and environment-friendly approach to lessen our need for water. Utilizing less water likewise puts less weight on our sewage treatment facilities which use an ample amount of energy for heating water.

For the past 50 years, the extraction of fresh water from icebergs has expanded by three folds. This is because of progression in the way of life a more significant amount of water is a need. Which likewise implies a growth in the interest for the power supply so with water.

Conserving water can likewise make the life of your septic system longer. This is by lessening soil immersion, and reducing any contamination because of leaks. Overloading municipal sewer systems can also make untreated sewage flow to lakes and rivers. The smaller the amount of water coursing through these systems, the lower the probability of contamination. Even the few groups like the community-wide domestic water preservation avoided the expensive sewage system development.

Saving Water Really is Important

Since safe and clean water is limited, people who have access to fresh water. They can take steps to control their consumption of water to avoid waste and shortage. We know that the planet is mostly covered with saltwater. And can only be consumed after undergoing a desalination process, which is quite expensive. Saving water really means a lot not only to humans but to all the species on Earth.

Events such as droughts further limit access to clean and fresh water. This means that people need to take extra steps to reduce water use and save as much water as possible. In some areas of the world, access to water is limited due to contamination.

Water is Life

Everything on Earth requires water to sustain itself. But abusing it means that we are reducing its ability to provide us with this basic necessity. Water is a limited resource. While Earth is a self-contained ecosystem, the planet always has, and will always have, the same amount of water. The population growth puts a strain on water supplies. And clean water is reduced by the pollution and contamination the human kind creates.

Clearly, people are particularly reducing the water supply due to pollution. So as other contaminants. On top of that, we are polluting the water for all of Earth's creatures, sending chemicals like oil and fertilizers through the rivers. That ultimately end up in the ocean.

Without fresh water, one will die in just a short span of time. It is a simple yet morbid fact that helps drive the point across, water is life. Water conservation is the potential, the most cost-effective and environmentally sound way to reduce the demand for water.

The Why and How of Water Conservation

Using the limited water supply wisely and caring for it properly are just a few of the many keys to conserve water. Remember that we have limited availability of water supply. This means that we do not have an endless amount of water. Keep in mind that it is our responsibility to understand and learn more about water conservation. Even so find ways to help keep the resources pure and safe for the coming generations.

Saving Water Saves Energy

A lot of energy is required to treat water and supply it to your home. The same as a tremendous amount of water expected to cool the power plants that produce electricity.

At home, heating water for showers, shaving, cooking and cleaning likewise uses a lot of energy.

That is the reason it's imperative to recollect to save energy and water in your home. We tend to have longer, hotter showers as the climate gets colder. By placing a water efficient shower and lessening the time spent in the shower you can spare energy and water.

One of the best means to save energy across the region and in your own house is to use water all the more effective.

Saving Water Saves Money

For instance, you have your own well and septic system, the extra gallons of water released each day will soak the soil in the septic system absorption field to a point where extensive repair or replacement is necessary.

Conserving water can extend the life of the system and delay the need for repair. If you live in an area serviced by a municipal water system, the greater your water use, the more you pay for water.

Also, water conservation can help prevent water pollution. Overloading a septic system may cause nutrient and bacterial contamination. Of nearby lakes, streams and drinking water, even the water from your own well. The smaller the amount of water flowing through these systems, the lower the likelihood of pollution.

Pollution costs money, too. Excessive weed growth in a lake caused by mineral enrichment from poorly functioning septic systems often means costly weed control measures paid for by you and your neighbors. Polluted home water wells, if they can be repaired at all, can cost thousands of dollars to fix.

Saving Water Saves Nature

Saving water likewise decreases the risk of natural disasters such as droughts. We have to reuse water in the same number of were likely to save a more significant amount of it.

Saving water turns out to be critical for up and coming generations. They won't have enough accessibility of water unless we wind up worried from this day at present.

We have to save water for plants as well. Earth's oxygen and a large portion of the food originate from plants. Plants require water for survival as well.

As the world modernizes, a greater amount of water is to be utilized as a part of beautifying urban communities and for the recreational reason. We have to consider it too.

What Can Be Done

We have first to understand that the preservation of water is the obligation of each person. It is to be done as a whole. No government authority or institution can help us in saving water unless we ourselves desire to. Right now is high time to do so.

Reducing the use of water reduces the energy required to process and deliver it to homes, business, farms, and communities. Which in turn helps to reduce pollution and conserve fuel resources.

Cutting off wastage of water will enable us to keep up the artistry of a city. Additionally, protecting our natural ecosystems from further damage is critical especially for the survival of some endangered species. The great pacific garbage patch is a great example of the worst side of our wasteful practices.

There are many efficient approaches to preserve water in and around your home. Look through this rundown for ways that will work for you.

Here are demonstrated means to conserve more water:

In the Kitchen

- Whenever you wash the dishes by hand—it is the most ideal way—don't leave the water running for rinsing. If you have two sinks, fill one with rinse water. If you just have one sink, use a shower gadget as opposed to giving the water a chance to run. This saves 200 to 500 gallons every month.
- When washing the dishes by hand, use a minimal quantity of detergent as much as possible. This limits the rinse water needed. This saves 50 to 150 gallons every month.
- Make sure not to defrost frozen foods with running water. Either prepare in advance by putting frozen things in the icebox or refrigerator overnight. Or defrost them in the microwave. This saves 50 to 150 gallons every month.
- Make sure not to run the faucet while you clean vegetables. Wash them in a filled sink or container. This saves 150 to 250 gallons every month.
- Keep a container or bottle of drinking water in the fridge. This beats the inefficient tendency for running tap water to cool it for drinking. This saves 200 to 300 gallons every month.
- Cook foods in less water, this keeps the nutrients value of the food as well.

In the Bathroom

- Whenever you are taking a shower, don't misuse the cold water while sitting tight for hot water to reach the showerhead. In a container/ pail, catch that water to use on watering the plants outside, or to flush in your toilet. This saves 200 to 300 gallons every month.
- Examine the toilet for leaks: Place dye tablets or food coloring into the tank. If the color shows up in the bowl without flushing, there is certainly a leak that ought to be repaired. This saves 400 gallons per month.
- Make doubly sure your toilet is an ultra-low volume flush model which utilizes only one and a half gallons each flush.
- Make sure to turn off the water or faucet while brushing your teeth: This saves three gallons every day.
- Don't use the toilet as an ashtray or wastebasket: Every time you flush a cigarette butt, facial tissue, or other small bits of trash, you're wasting gallons of water. Put them in the proper garbage bin.
- Take shorter showers: One way to cut down on water use is to turn off the shower after soaping up, then turn it back on to rinse. A four-minute shower uses approximately 20 to 40 gallons of water.

In the Yard or Garden

- Avoid watering the lawn on windy days: There's excessive evaporation. This can waste up to 300 gallons in a single watering.
- It is better to water during the cool parts of the day: Preferably in the early morning to keep from the development of an organism. This saves 300 gallons.
- In case you have a pool, utilize a pool cover to eliminate evaporation: It will likewise keep your pool cleaner and lessen the need to include chemicals. This saves 1,000 gallons every month.
- Using a pail of soapy water, clean the car and make sure to drive it onto a lawn: The water used can help water the grass at the same time. Only use the hose for rinsing - this easy practice can save as much as 150 gallons when washing a car.
- Minimize watering on cool and cloudy days and do not water in the rain: Change or deactivate automatic sprinklers. This can save up to 300 gallons each time.
- Put a layer of mulch around trees and plants: Pieces of bark, peat greenery or rock gradually slow down evaporation. This saves 750 to 1,500 gallons every month.
- Tell your kids not to play with the garden hose: This saves 10 gallons every moment.
- Whenever you allow your kids to play in the sprinklers, ensure it's just when you're watering the yard If it's not very cool around that time of day.

Saving Water in Special Conditions

At some point, it is necessary to use extra measures to reduce the amount of water you consume at home. Although suitable for any kind of situation, these techniques may be especially helpful. When water levels are high around your house, or your community water system temporarily loses the capacity to supply adequate amounts of water. You should consider these changes:

Use much around trees and shrubs and in garden beds. This greatly reduces the amount of water lost through evaporation. And so reduces the need for watering.

Consider using a drip irrigation system in your garden. It supplies water only to the root zones of plants, and reduces weeding because it doesn't water areas between crop rows and hills.

Use only plant varieties that are well adapted to your locality and soil conditions. Less suitable varieties may need more fertilizer or water to live.

Use the water from your roof downspouts for watering your garden and flower beds.

Strategies to Support Water Conservation

Conserving water for individual use in urban areas (counting use by families and districts) needs an inspection. Both the supply and demand for water.

A part of the strategies that can aid water preservation activities. And handle the water shortage issue include:

Rainwater Harvesting

Rainwater harvesting is essentially a technique to store water and get it. This is for fair usage on the later day and period. The system has unique units which incorporate transportation of rainwater, filtration and storing of the processed water.

It will be more beneficial to install a rainwater storing unit in our homes to spare more water.

- Rooftop rainwater harvesting.
- Revival of traditional water harvesting structures.
- Micro-catchment water harvesting.
- Recharge structures for wells and bore wells.

Sustainable Water Usage

Sustainable water supply includes an arrangement of joined activities and not disconnected strategies. It relies upon the person's ability to save water, administrative regulations, and changes in the building industry, industrial forms production, land occupation, and so forth. The challenge is to make components of direction. How reasonable it is to guarantee the sustainability of the system.

- Minimize domestic water consumption.
- Recycling of wastewater.
- Improved irrigation methods.

Encourage natural regeneration of vegetation and supplementing with artificial regeneration.

Regeneration is 'the renewal of a forest crop by natural or artificial means.

Using crops by sowing, planting and may it be through artificial methods. These have a greater impact on conserving water. The natural and artificial regeneration of vegetation is a dynamic procedure. By which life recolonizes land when the vegetation has been somewhat or completely devastated. Life restores the lost ground through the instrument of the progression of species.

Maintain and Improve the Quality of Water

Water quality in a waterway impacts the means in which communities use the water for activities like drinking, swimming or business purposes. All the more particularly, the water might be utilized by the group for the production of edible fish, shellfish and crustaceans, protection of aquatic ecosystems, wildlife habitats, supplying drinking water and the like.

- Collection and treatment of wastewater effluents.
- Pollution check.

Raising Awareness of Water Conservation

We all need to go hand in hand together because water is a global priority. And it is imperative to save as much water we can in any way possible. If it is not for us, then for the generations to come and if not for the generation, for the world that we live in, the Earth. Building awareness seems simple, yet so hard to deliver to the human race. One thing for sure is the easiest, to start it in our home, with ourselves. Rather than sitting tight for another person to begin conserving, let us, as an individual, initiate in conserving.

Most importantly, we have to educate everyone about how essential water really is. That is the very least way we can save water. The more we educate people, all the more water we save. Every leakage ought to be fixed in the drainage system wherever available.

Wasting water has turned into a powerful environmental issue - both at consumer and industrial levels. It has turned out to be essential for people and organizations alike to discover approaches to decrease the wastage of water and conserve it.

There are various approaches to save water. Conserving is one and reusing it is another. Given that we live in a zone lying down to dry season, it just shows well to save each. Means we can and using water twice is one awesome approach to extend this valuable resource further.

Extend your Water Conservation Measures

- Recycle your water wherever, whenever you can. Collect the cold water you run before it's hot enough to shower and use it to water plants or flush the toilet. After, rinse water from dishes and food preparation can be collected and used to soak other dishes.

- Insulate your water pipes. It's convenient and low-cost to insulate your water pipes with pre-slit foam pipe insulation. You'll get hot water faster plus avoid wasting water while it heats up.

- Eat less water-intensive foods. Our diets regime explains for about half of all the water we consume. All food has a water footprint, but some are much larger than others. Eating less beef, one of the most water-intensive foods, is a smart place to start. Shifting away from animal products to a plant-based diet can shrink your water footprint.

- Buy less. Consumer products are an often-overlooked source of water use, accounting for up to a third of most people's water footprint. Buying less of everything. From clothing to electronics to household goods can dramatically decrease your water footprint.

Water Facts and Trivia

- The overall number of water has continued as is for two billion years on our planet.
- Around 39,000 gallons of water is expected to create a vehicle.
- Around a billion people need stable access to clean water.
- It takes around 6 gallons of water to cultivate a sole portion of lettuce.
- More than 2,600 gallons are needed to deliver a single serving of steak.
- A typical shower utilizes around 25 gallons of water.
- Brushing your teeth utilizes around 10 gallons of water.
- Bathtub utilizes roughly 36 gallons of water.
- Shaving utilizes roughly 20 gallons of water.
- Dishwashing utilizes roughly 30 gallons of water.
- An automatic dishwasher utilizes roughly 16 gallons of water for every cycle.
- Washing your hands utilizes roughly 2 gallons of water.
- Flushing the toilet utilizes 5-7 gallons for each flush.
- A typical washing machine cycle utilizes 60 gallons of water.
- Watering outside utilizes around 10 gallons for every moment.

WATER CONSERVATION TECHNIQUES

The following methods would prove useful for it:

Protection of Water from Pollution

If the total fresh water available on the earth remains pollution free, it is sufficient to meet the drinking water needs of the existing population of the world, unfortunately a large portion of fresh water does not remain fit for use of the living world due to increasing economic activities, urbanization etc.

Oceanic water in the form of ecological system of seas is an important environmental system, but during the last century pollution has spread in large proportions.

Surface water is mainly found in rivers and lakes and underground water is found under land at different depths, but which have become polluted.

Large cities located on banks of rivers are directly disposing off different wastes without treatment in rivers. Similarly, tourism has spread pollution at war speed on famous lakes and sea coasts. Man is greatly dependent on groundwater for his water related necessities, but some special industrial units have also polluted this amount of water stored in the security cover of the ground.

All the available water store on the earth should be kept pollution-free because pressure of demand for water is increasing on a large part of the earth. Normally, it is presumed that sea water being saline, is not fit to be used by human beings, but indirectly it is useful for maintaining living organisms with whom man is related, e.g., fish provide nutrition to a large portion of the world.

Hence, it is necessary to have a cooperative policy at the international level for control of oily and radioactive pollution of the oceans. On 18th November, 2002, an oil tanker of Bahama met with an accident and drowned in the sea near the 'Coast of Death' sea area 233 km from north-west coast of Spain. 77,000 metric tones of oil poured out from this tanker and spread over the nearby seas and polluted the sea water. During the Gulf War of 2003 also, the water of Red Sea, Gulf Persian and Mediterranean Sea adjacent to the Arabian Sea became polluted.

The origin of water crisis initially started due to its pollution and qualitative deterioration, which became acute due to increasing demand for it. For preventing water pollution, important water sources should not be made places for disposal of wastes. Industrial units should dispose off water only after its treatment. Bathing and other such activities should be prohibited near drinking water sources.

Weeds produced in water should be controlled. Water should become re-usable after physical, chemical, mechanical and organic processing methods. Rivers and lakes should not be used as canals for transport of goods as is being done in Rhine River of Europe, Volga river of Russia, and the Ganges, Yamuna and Kaveri (Cauvery) rivers in India. Thus, the most important aspect of water conservation is control on water pollution.

Redistribution of Water

Water found on the surface of the earth is not equally distributed. Existing form of distribution also becomes a reason for the water crisis. In the African continent, though there is maximum hydro electric production because of excess availability of water in Mediterranean regions, but the Sahara desert situated in the north of Africa and the greatest desert of the world, suffers from water crisis for the whole year.

Droughts faced by the Sahel region in the north of Sahara desert are world famous. Similarly, there is maximum rainfall of the world (1,187 cm. in Mawsynram) in north-east India, whereas there is only 50 cm. rainfall in the west. As a result of it, more than 60 per cent portion of water of Brahmaputra and its companion rivers in the north east flows to the saline seas uselessly, whereas rivers of western Rajasthan remain dry for most of the time in a year.

Hence, by arranging supply of water from areas having lesser demand to the areas having greater demand, water crisis can be minimized. By construction of surface water reservoirs and storage of excess water in them, supply can be made to scarcity affected areas. This work can be accomplished by development of water reservoirs and canal network.

Excess rainfall water which flows away from rivers without being used, can be stored by construction of water reservoirs, from where it can be supplied for agriculture, industries, urban areas etc. Facilities of fisheries and transport also exist in stored water. These reservoirs are also constructed for protection from floods, and apart from flood protection, such water can also be used for different purposes.

Redistribution of water is also possible through canal system. Canal system transfers water from excess rain water areas to scarce rain water areas and conserves water for different uses. Indira Gandhi Canal is such a type of canal system which has brought water of Himalayas to western Rajasthan and changed the arid ecology.

Its length was decided as 2640 km in the beginning, which had to link river Ganges of north India with river Kaveri known as 'Ganges of the South'. It was also proposed to include central rivers Narmada, Tapti, Godavari, Krishna and Pensnar falling in the way. Former President of India A.P.J. Abdul Kalam took a lead in 2003 for its implementation and its work is being started by giving a modern shape to it. It will be able to transfer water from excess rainfall water areas to scarce rainfall areas every year.

Thus, by developing artificial canal system through artificial rivers, excess water can be conserved. For this, canal system should be developed from national level to micro level. Along with big rivers, canal system should also be developed by the side of small rivers to conserve large amounts of water. Though development of canal system will have to face geographical, economic, social, cultural (rehabilitation) and political problems, but its future results would be favourable from the geographical point of view.

Rational use of Groundwater

Groundwater meets 25 per cent of total supply of water in the world, remaining 75 per cent supply is met by surface water sources of rivers, lakes etc. Demand for groundwater goes on increasing in proportion to its available quantity due to which quantity of groundwater goes on decreasing. After exploitation of groundwater, its re-infiltration takes a very long time to complete.

Hence, groundwater exploitation should be only in proportion to its recharging capacity. In India, the maximum groundwater is utilized for agriculture. Instead of growing crops requiring water according to agricultural climatic conditions, crops of commercial importance requiring more water are grown, due to which overexploitation of groundwater is done.

In Rajasthan, out of 237 blocks, 203 were in the safety zone from the point of view of groundwater in 1984, but due to excessive over exploitation of groundwater during the last two decades, the number of safety blocks reduced to only 49 in the year 2001. Remaining blocks have gone to the 'Dark Zone', where groundwater is not available in the required quantity for different uses in future.

By making only conscientious exploitation of groundwater, it can be proportionately conserved. For that, crops requiring lesser water should be given priority over crops requiring more water. Lesser water should be used by industries. For recharging groundwater, necessary drainage area should also be made available.

Population Control

Water crisis has assumed dangerous proportions due to fast increase in population and deterioration in quantitative and qualitative aspects of water resources. Demand for water is continuously increasing. Along with population increase, demand for fresh water has increased due to expansion of industries and increasing urbanization.

Demand for water increased 35 times from the year 1700 to the decade of 2000. Demand for water has increased more in developing countries. For the year 2001, demand for quantity of water in the whole world was estimated to be 435 cubic kms. 60 per cent of this is required in agriculture, 30 per cent in industries and 10 per cent for cooking, bathing and drinking. Hence, demand for water can also be controlled through population control and at the same time, qualitative deterioration of water can also be prevented by population.

Population of the world would become around 800 crores by the year 2025. In this context, the United Nations Organization has warned that in case population control is not exercised, the whole world will have to face grave water crisis. United Nations Organization declared the year 2003 as the 'Fresh Water Year'.

Renovation of Traditional Water Sources

In India, traditional water storage places have been able to meet the demand of drinking water in many regions but they have been renovated from time to time. Water stored in traditional water sources has been used for both purposes, agriculture as well as for drinking.

Important traditional water conservation methods for irrigation include Kuhul in hilly areas, Jing (Ladakh), Kool, Water Kundis called Khoop in Arunachal Pradesh, Zabo method of Nagaland, Aabi tanks of Haryana, Dong Pokhar of Assam, Bandhare of Maharashtra, Kere of Karnataka, Iree (Tanks) in Tamil Nadu, jackwell of Andaman Nicobar and Nadi, Tanka, Kund, Khadeen, Kui, Bedi, Baori, Jhalra, Toba etc. of Rajasthan.

Beginning of traditional water conservation methods was made in Java (Jordan) in the year 3,000 B.C. by construction of a vast water reservoir. In India, a system of water conservation and drainage was found in Dhauliveera habitations 1000-1500 B.C.), which was constructed during the Harappan civilization.

These traditional water conservation methods had been developed in India looking to the nature of rainfall in different regions, but the increasing population necessitated extension of agriculture, leading to deterioration of these traditional sources. Conservation of any traditional water source, instead of being limited to the water reservoir portion, extends to the whole drainage area system where rain water is stored after flowing.

Due to increasing area of agriculture, drainage area of water has been destroyed resulting in reduction of water flow therein. This has caused crisis for the existence of traditional water sources. Water crisis can be prevented by renovating traditional water conservation sources. Unless complete conservation of rain water is done in any region, the dream of water conservation cannot be fulfilled. Hence, supervision has to be done regarding renovation of dying traditional water conservation places.

Now, the whole world has agreed with this fact that water available in its vast form also shall have to be conserved to keep it within our reach. From this point of view, ownership of traditional water sources should be at the village level as has also been agreed by the World Bank.

Use of Modern Irrigation Methods

At the world level, 69 per cent water is used for the agricultural sector. Necessary requirement of water in agriculture is met by surface water sources and groundwater.

Surface water used for agriculture is obtained through canals and tanks and underground water from wells and tube wells. Unfortunately, much water is wasted through traditional methods of irrigation. Canal irrigation does not distribute water in a balanced way and it creates problems of water logging.

A large portion of water can be conserved by adoption of modern methods of irrigation. Irrigation consumes double the quantity of water in comparison to all other uses. Sprinkler and drip irrigation methods save 50 per cent water.

In drop or drip irrigation method, pipes with holes are spread over the surface of land so that the crop directly receives water. There is no loss due to evaporation in this system and almost 95 per cent water is utilized. Thus, maximum water is utilized by this method. Improved modern irrigation methods are thus useful for conservation of water.

Increasing Forest Cover

According to hydrological movements, water is received through rainfall every year m different quantities on the surface of the earth. This water flows on the surface and reaches the seas. Some part of rainwater is stored in stable water reservoirs (lakes and tanks), whereas some quantity of water infiltrates into the land and takes the form of groundwater.

Due to increasing deforestation during the last century, most of the rainwater flowed away to the saline seas without infiltrating into the ground. Water crisis also developed during the last decade in Cherapunji, which gets highest rainfall in the world, because forest cover has been destroyed there due to mining of limestone. As a result of it, rain water flows away very fast to the rivers. A similar thing is happening in the Dehradun area of Uttaranchal.

The old tradition of tree plantation on the banks of rivers and tanks will have to be revived. Forest cover will have to be developed on uncultivable waste lands and hilly slopes on a large scale. Since trees bear drought conditions for a long duration as compared to crops, hence trees are helpful in reducing the demand for water along with recharging water sources.

Change in Crop Pattern

Excess water is not required if crops are grown according to agro-climatic conditions but in the present race of development, changing crop pattern with higher profits has replaced them. These commercial crops require more water than the traditional crops. In north-eastern part of Rajasthan, crops were not grown as per availability of water and intensive cultivation was adopted during the last three decades.

Groundwater was over exploited because of non-availability of surface water and plantation of commercial crops requiring more water. It created serious water crisis. Hence, keeping in mind the experience of Rajasthan, crop rotation should be adopted according to agro-climatic conditions. Agro forestry and horticulture should be given priority in areas having scarcity of water.

Flood Management

A large portion of fresh water in the world becomes devastating due to floods in India, out of a total land area of 32.8 crore hectares, in India, 4 crore hectare land is flood affected, out of which 3.2 crore hectare land can be protected from floods. By construction of embankments and canals a large part of land can be conserved besides minimizing flood losses.

Intensive afforestation can also provide security from floods. If will be helpful in absorption of water in the soil. Drainage areas of Ganges, Yamuna, Mahanadi, Damodar, Kosi and other rivers have been taken up in flood management and security to some limit has been provided to the 1.44 crore hectare land.

Use of Geothermal Water

Water is also received from hot waterfalls regularly at many places on earth. Scarcity of water can be met to a certain extent by using such geothermal water.

Conserving Water in Industries

About 23 per cent of total fresh water available on the earth is used by industries the world over. Some special industries consume water to a large extent, whereas some industries pollute the major portion of water. Dyeing industry and leather industry are such type of industries which pollute the water.

To manufacture one ton of steel, 300 tons of water is required. Consumption of water in industries is done in quantitative as well as qualitative forms. Proportionate consumption of water in industries in developed' countries is more (50%), out of which 75 per cent demand is met from surface water sources and 25 per cent from groundwater sources.

Water has to be both protected from pollution in industries, and re-used after processing. Recycling of water should be developed because normally industrial units dispose off water on the ground surface after using it only once. Because of not re-using it after processing, it also pollutes other water sources. Demand for water can also be reduced in industries by recycling used water.

Reuse of Urban Waste

Demand for water has increased in cities due to increasing urbanization. There is no provision for waste water treatment in many big towns and cities of various countries of the world. Instead of being reused, it pollutes other water sources.

Such condition is visible in cities of Delhi, Agra and Mathura on the banks of river Yamuna, whereas in many countries, urban water is used after treatment in nearby fields for growing vegetables and fruits. After use of water in urban areas, disposed waste water can be treated and conserved for use in agriculture in the peripheral areas of cities. Such policies should be incorporated while planning urban development.

Water Conservation by Municipal Bodies

Municipal bodies should manage both individual demand and supply of water as well as conserve water. Municipal laws should provide for collection of rain water from roof tops and implement it. Individual awareness is very important in water conservation.

Every individual should develop the attitude of water conservation and prevent every drop of water from being wasted. As far as possible, one should conserve rain water along with maintaining quality of the water. Depth of traditional water sources should be maintained by controlling its cleanliness.

Since the maximum part of water is exploited in agriculture, hence water should be conserved in agriculture in different forms. Cultivating fields in off season helps in maintaining soil moisture. If cultivation is done up to 30 cm. depth, moisture can go up to 90 cm. depth, and capillary action prevents moisture coming out from the soil.

Besides this, soil moisture can also be maintained by before-time sowing, proper use of fertilizers, pesticides and Weedicides. Green manure and crop rotation should also be adopted.

Environment balancing is the main basis of water conservation. Changes coming in the world environment result in shortage in the quantity of fresh water. Due to rise in temperature of the world, fresh water in the form of snow is melting and drifting towards saline oceans. This change is visible from Antarctica to Gangotri Glacier in the Himalayas.

Change in climate has also caused change in nature and quantity of rainfall, due to which existing unevenness in distribution of water has also increased. Thus, the mentality of World Environment Balancing has to be developed at the individual angle and conservation has to be worked on each stage.

The objective behind the constitution of 'World Water Commission' at the Hague in 1999 was to consider questions related to water conservation and creating an environment so that strategy of water management may be worked out to search out solutions to the impending water crisis. Five groups have been constituted here.

First, for change in information system; second related to energy; third for considering issues related to organic technology; fourth to discuss institutional changes and the fifth panel is for consolidating serially future forecasts based on study of statistics. Apart from them, there is also a panel for traditional areas to consider water for food, water for human consumption and water for environment.

RAIN WATER HARVESTING

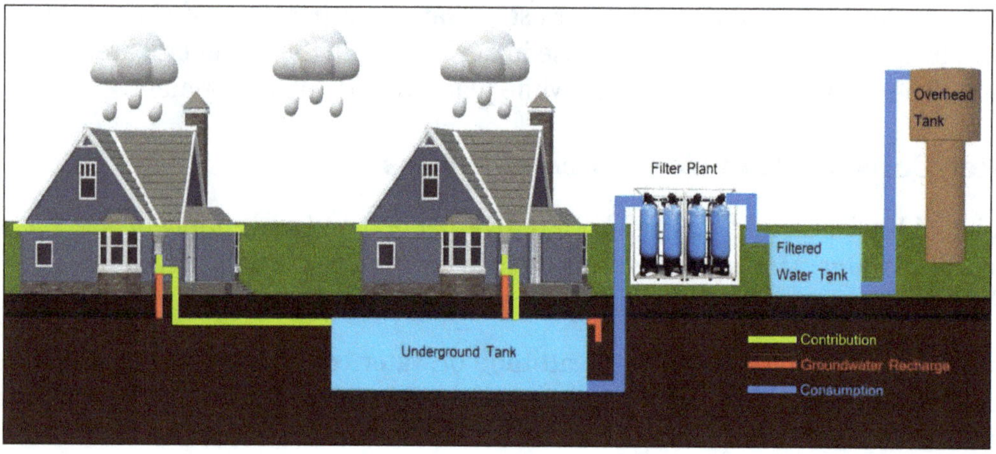

In the present scenario management and distribution of water has become centralized. People depend on government system, which has resulted in disruption of community participation in water management and collapse of traditional water harvesting system.

As the water crisis continues to become severe, there is a dire need of reform in water management system and revival of traditional systems. Scientific and technological studies needs to be carried out to assess present status so as to suggest suitable mitigative measures for the revival to traditional system/wisdom. Revival process should necessarily be backed by people's initiative and active public participation.

Living creatures of the universe are made of five basic elements, viz., earth, water, fire, air and sky. Obviously, water is one of the most important elements and no creature can survive without it. Despite having a great regard for water, we seem to have failed to address this sector seriously. Human being could not save and conserve water and it sources, probably because of its availability in abundance. But this irresponsible attitude resulted in deterioration of water bodies with respect to quantity and quality both. Now, situation has arrived when even a single drop of water matters. However. " better late than never", we have not realized the seriousness of this issue and initiated efforts to overcome those problems.

System of collection rainwater and conserving for future needs has traditionally been practiced in India. The traditional systems were time-tested wisdom of not only appropriate technology of Rainwater Harvesting, but also water management systems, where conservation of water was the prime concern. Traditional water harvesting systems were Bawaries, step wells, jhiries, lakes, tanks etc. These were the water storage bodies to domestic and irrigation demands. People were themselves responsible for maintenance to water sources and optimal use of water that could fulfill their needs.

The term rainwater harvesting is being frequently used these days, however, the concept of water harvesting is not new for India. Water harvesting techniques had been evolved and developed centuries ago.

Ground Water

Ground water resource gets naturally recharged through percolation. But due to indiscriminate development and rapid urbainzation, exposed surface for soil has been reduced drastically with resultant reduction in percolation of rainwater, thereby depleting ground water resource. Rainwater harvesting is the process of augmenting the natural filtration of rainwater in to the underground formation by some artificial methods. "Conscious collection and storage of rainwater to cater to demands of water, for drinking, domestic purpose & irrigation is termed as Rainwater Harvesting."

Why Harvest Rainwater?

This is perhaps one of the most frequently asked question, as to why one should harvest rainwater. There are many reasons but following are some of the important ones.

- To arrest ground water decline and augment ground water table.
- To beneficiate water quality in aquifers.
- To conserve surface water runoff during monsoon.
- To reduce soil erosion.
- To inculcate a culture of water conservation.

How to harvest rainwater:

Broadly there are two ways of harvesting rainwater:

- Surface runoff harvesting.
- Roof top rainwater harvesting.

Surface Runoff Harvesting

In urban area rainwater flows away as surface runoff. This runoff could be caught and used for recharging aquifers by adopting appropriate methods.

Roof top Rainwater Harvesting (RTRWH)

It is a system of catching rainwater where it falls. In rooftop harvesting, the roof becomes the catchments, and the rainwater is collected from the roof of the house/building. It can either be stored in a tank or diverted to artificial recharge system. This method is less expensive and very effective and if implemented properly helps in augmenting the ground water level of the area.

Components of the Roof Top Rainwater Harvesting System

The illustrative design of the basic components of roof top rainwater harvesting system is given in the following typical schematic diagram.

The system mainly constitutes of following sub components:

- Catchment.
- Transportation.
- First flush.
- Filter.

The surface that receives rainfall directly is the catchment of rainwater harvesting system. It may be terrace, courtyard, or paved or unpaved open ground. The terrace may be flat RCC/stone roof or sloping roof. Therefore the catchment is the area, which actually contributes rainwater to the harvesting system.

Transportation

Rainwater from rooftop should be carried through down take water pipes or drains to storage/harvesting system. Water pipes should be UV resistant (ISI HDPE/PVC pipes) of required capacity. Water from sloping roofs could be caught through gutters and down take pipe. At terraces, mouth of the each drain should have wire mesh to restrict floating material.

First Flush

First flush is a device used to flush off the water received in first shower. The first shower of rains needs to be flushed-off to avoid contaminating storable/rechargeable water by the probable contaminants of the atmosphere and the catchment roof. It will also help in cleaning of silt and other material deposited on roof during dry seasons Provisions of first rain separator should be made at outlet of each drainpipe.

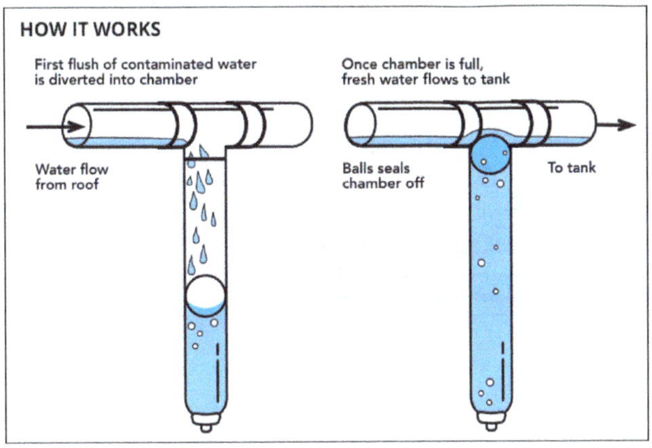

Benefits

- Prevents sediment, bird droppings, spider, insects, mosquito eggs and debris from entering the rainwater tank.
- Improves water quality, protects pumps and internal appliances.
- Ideal to use in conjuction with a rain head.
- Easy installation, just add pipe and glue.
- No mechanical parts.

Filter

There is always some skepticism regarding Roof Top Rainwater Harvesting since doubts are raised that rainwater may contaminate groundwater. There is remote possibility of this fear coming true if proper filter mechanism is not adopted. Secondly all care must be taken to see that underground sewer drains are not punctured and no leakage is taking place in close vicinity. Filters are used fro treatment of water to effectively remove turbidity, colour and microorganisms. After first flushing of rainfall, water should pass through filters. There are different types of filters in practice, but basic function is to purify water.

Sand Gravel Filter

These are commonly used filters, constructed by brick masonry and filleted by pebbles, gravel, and sand as shown in the figure. Each layer should be separated by wire mesh.

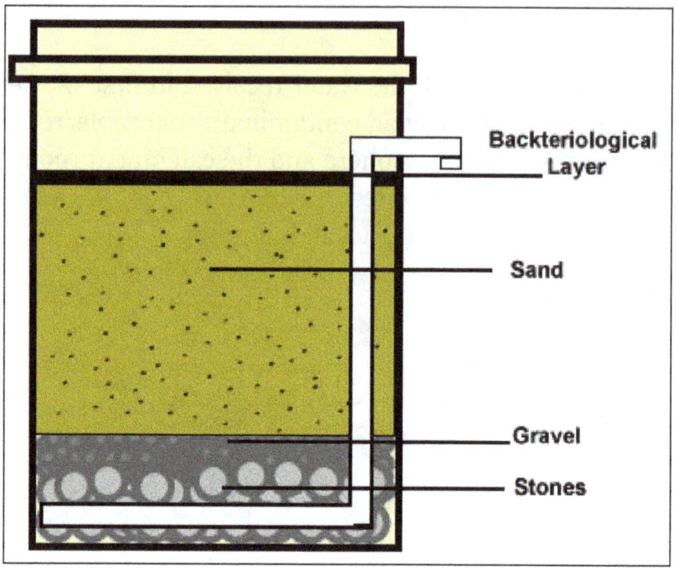

Charcoal Filter

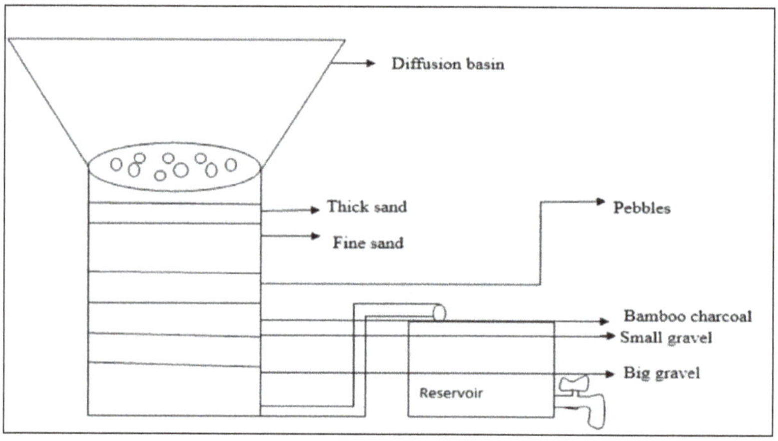

Charcoal filter can be made in-situ or in a drum. Pebbles, gravel, sand and charcoal as shown in the figure should fill the drum or chamber. Each layer should be separated by wire mesh. Thin layer of charcoal is used to absorb odor if any.

PVC - Pipe Filter

This filter can be made by PVC pipe of 1 to 1.20 m length; Diameter of pipe depends on the area of roof. Six inches dia. pipe is enough for a 1500 Sq. Ft. roof and 8 inches dia. pipe should be used for roofs more then 1500 Sq. Ft. Pipe is divided into three compartments by wire mesh. Each component should be filled with gravel and sand alternatively as shown in the figure. A layer of charcoal could also be inserted between two layers. Both ends of filter should have reduce of required size to connect inlet and outlet. This filter could be placed horizontally or vertically in the system.

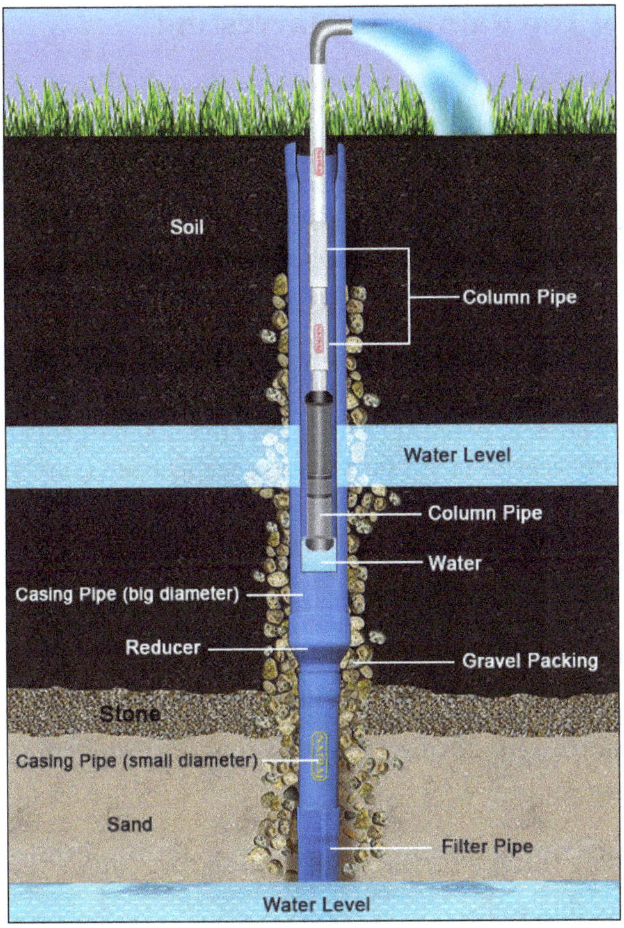

Sponge Filter

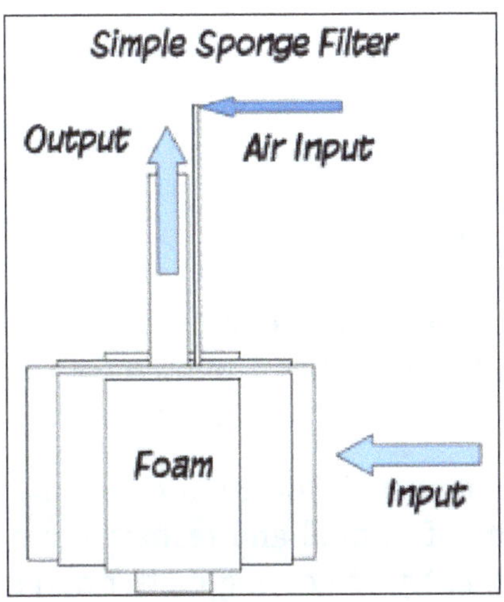

It is a simple filter made from PVC drum having a layer of sponge in the middle of drum. It is the easiest and cheapest form filter, suitable for residential units.

Methods of Roof Top Rainwater Harvesting

Storage of Direct Use

In this method rain water collected from the roof of the building is diverted to a storage tank. The storage tank has to be designed according to the water requirements, rainfall and catchment availability. Each drainpipe should have mesh filter at mouth and first flush device followed by filtration system before connecting to the storage tank. It is advisable that each tank should have excess water over flow system.

Excess water could be diverted to recharge system. Water from storage tank can be used for secondary purposes such as washing and gardening etc. This is the most cost effective way of rainwater harvesting. The main advantage of collecting and using the rainwater during rainy season is not only to save water from conventional sources, but also to save energy incurred on transportation and distribution of water at the door-step. This also conserve groundwater, if it is being extracted to meet the demand when rains are on.

Recharging Ground Water Aquifers

Ground water aquifers can be recharged by various kinds of structures to ensure percolation of rainwater in the ground instead of draining away from the surface. Commonly used recharging methods are:-

- Recharging of bore wells.
- Recharging of dug wells.
- Recharge pits.
- Recharge Trenches.
- Soak ways or Recharge Shafts.
- Percolation Tanks.

Recharging of Bore Wells

Rainwater collected from rooftop of the building is diverted through drainpipes to settlement or filtration tank. After settlement filtered water is diverted to bore wells to recharge deep aquifers. Abandoned bore wells can also be used for recharge.

Optimum capacity of settlement tank/filtration tank can be designed on the basis of area of catchement, intensity of rainfall and recharge rate as discussed in design parameters. While recharging, entry of floating matter and silt should be restricted because it may clog the recharge structure. "first one or two shower should be flushed out through rain separator to avoid contamination. This is very important, and all care should be taken to ensure that this has been done."

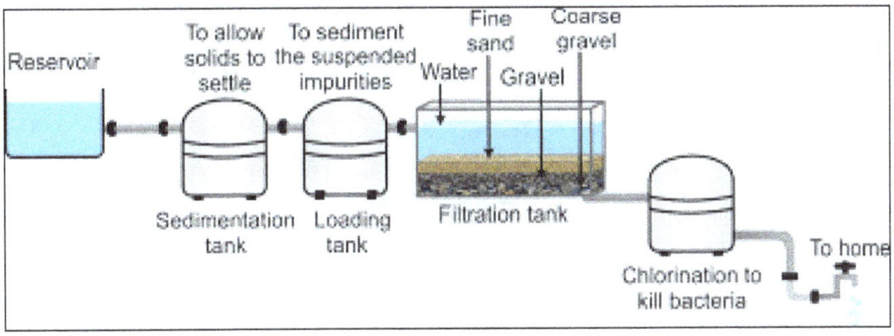

Recharge Pits

Recharge pits are small pits of any shape rectangular, square or circular, contracted with brick or stone masonry wall with weep hole at regular intervals. to of pit can be covered with perforated covers. Bottom of pit should be filled with filter media.

The capacity of the pit can be designed on the basis of catchment area, rainfall intensity and recharge rate of soil. Usually the dimensions of the pit may be of 1 to 2 m width and 2 to 3 m deep depending on the depth of pervious strata. These pits are suitable for recharging of shallow aquifers, and small houses.

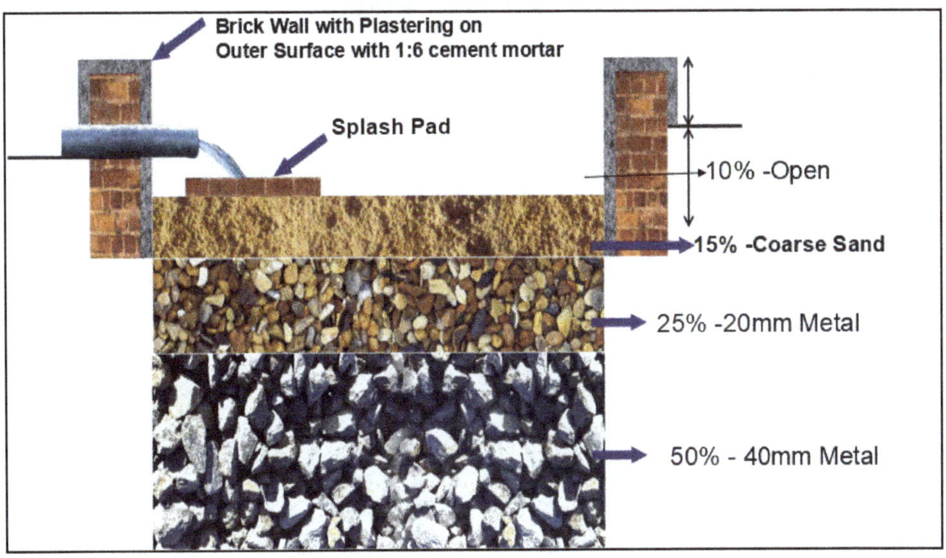

Soak away or Recharge Shafts

Soak away or recharge shafts are provided where upper layer of soil is alluvial or less pervious. These are bored hole of 30 cm dia. up to 10 to 15 m deep, depending on depth of pervious layer. Bore should be lined with slotted/perforated PVC/MS pipe to prevent collapse of the vertical sides. At the top of soak away required size sump is constructed to retain runoff before the filters through soak away. Sump should be filled with filter media.

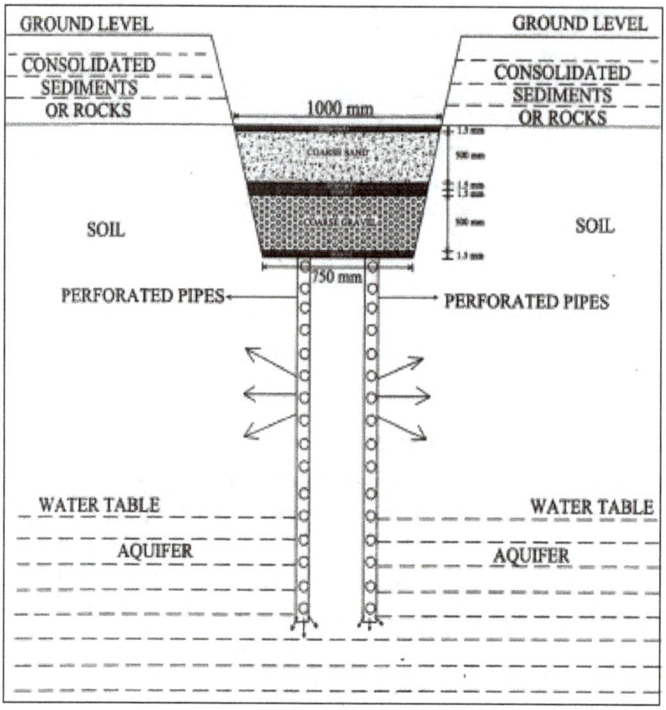

Vertical Recharge Shaft

Recharging of Dug Wells

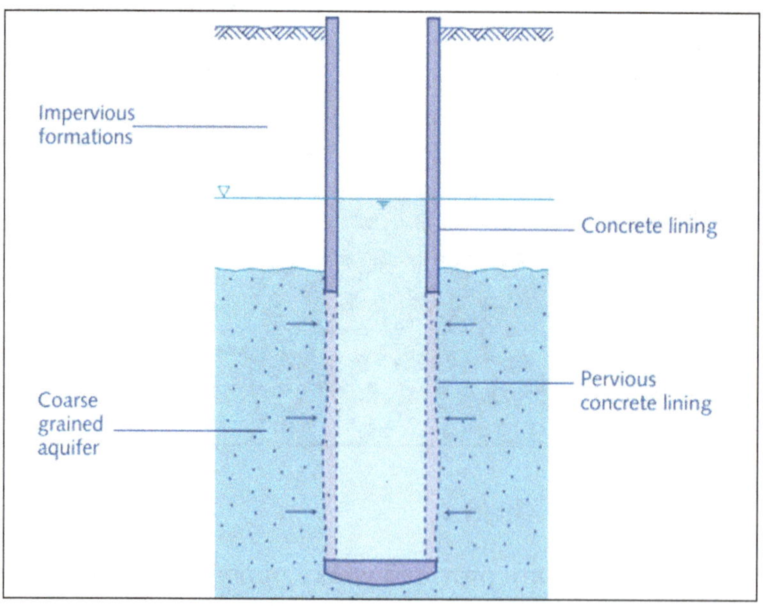

Dug well can be used as recharge structure. Rainwater from the rooftop is diverted to dug wells after passing it through filtration bed. Cleaning and desalting of dug well should be done regularly to enhance the recharge rate. The filtration method suggested for bore well recharging could be used.

Recharge Trenches

Percolation Tanks

Percolation tanks are artificially created surface water bodies, submerging a land area with adequate permeability to facilitate sufficient percolation to recharge the ground water. These can be built in big campuses where land is available and topography is suitable.

Surface run-off and roof top water can be diverted to this tank. Water accumulating in the tank percolates in the solid to augment the ground water. The stored water can be used directly for gardening and raw use. Percolation tanks should be built in gardens, open spaces and roadside green belts of urban area.

Do's and Don'ts

Harvested rainwater is used for direct usage or for recharging aquifers. It is most important to ensure that the rainwater caught is free from pollutants. Following precautionary measures should be taken while harvesting rainwater:

- Roof or terraces uses for harvesting should be clean, free from dust, algal plants etc.
- Roof should not be painted since most paints contain toxic substances and may peel off.
- Do not store chemicals, rusting iron, manure or detergent on the roof.
- Nesting of birds on the roof should be prevented.
- Terraces should not be used for toilets either by human beings or by pets.
- Provide gratings at mouth of each drainpipe on terraces to trap leaves debris and floating materials.

- Provision of first rain separator should be made to flush off first rains.
- Do not use polluted water to recharge ground water.
- Ground water should only be recharged by rainwater.
- Before recharging, suitable arrangements of filtering should be provided.
- Filter media should be cleaned before every monsoon season.
- During rainy season, the whole system (roof catchment, pipes, screens, first flush, filters, tanks) should be checked before and after each rain and preferably cleaned after every dry period exceeding a month.
- At the end of the dry season and just before the first shower of rain is anticipated, the storage tank should be scrubbed and flushed off all sediments and debris.

DESALINATION

Desalination is a process that takes away mineral components from saline water. More generally, desalination refers to the removal of salts and minerals from a target substance, as in soil desalination, which is an issue for agriculture.

Saltwater is desalinated to produce water suitable for human consumption or irrigation. One by-product of desalination is salt. Desalination is used on many seagoing ships and submarines. Most of the modern interest in desalination is focused on cost-effective provision of fresh water for human use. Along with recycled wastewater, it is one of the few rainfall-independent water sources.

Due to its energy consumption, desalinating sea water is generally more costly than fresh water from rivers or groundwater, water recycling and water conservation. However, these alternatives are not always available and depletion of reserves is a critical problem worldwide. Desalination processes are usually driven by either thermal (e.g. distillation) or electrical (e.g., photovoltaic or wind power) as the primary energy types.

Currently, approximately 1% of the world's population is dependent on desalinated water to meet daily needs, but the UN expects that 14% of the world's population will encounter water scarcity by 2025. Desalination is particularly relevant in dry countries such as Australia, which traditionally have relied on collecting rainfall behind dams for water.

According to the International Desalination Association, in June 2015, 18,426 desalination plants operated worldwide, producing 86.8 million cubic meters per day, providing water for 300 million people. This number increased from 78.4 million cubic meters in 2013, a 10.71% increase in 2 years. The single largest desalination project is Ras Al-Khair in Saudi Arabia, which produced 1,025,000 cubic meters per day in 2014. Kuwait produces a higher proportion of its water than any other country, totaling 100% of its water use.

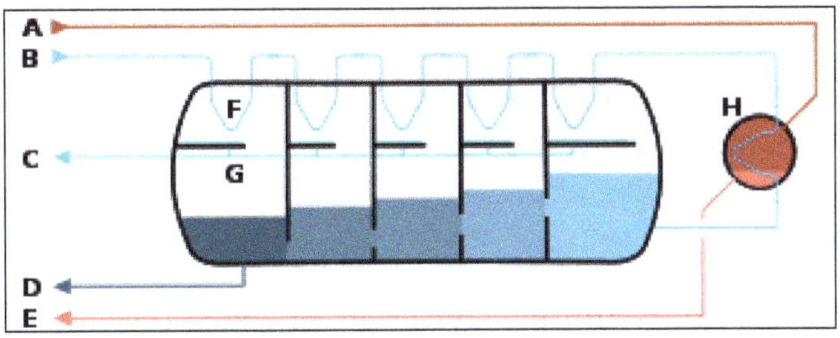

Schematic of a multistage flash desalinator A – steam in B – seawater in C – potable water out D – brine out (waste) E – condensate out F – heat exchange G – condensation collection (desalinated water) H – brine heater The pressure vessel acts as a countercurrent heat exchanger. A vacuum pump lowers the pressure in the vessel to facilitate the evaporation of the heated sea water (brine) which enters the vessel from the right side (darker shades indicate lower temperature). The steam condensates on the pipes on top of the vessel in which the fresh sea water moves from the left to the right.

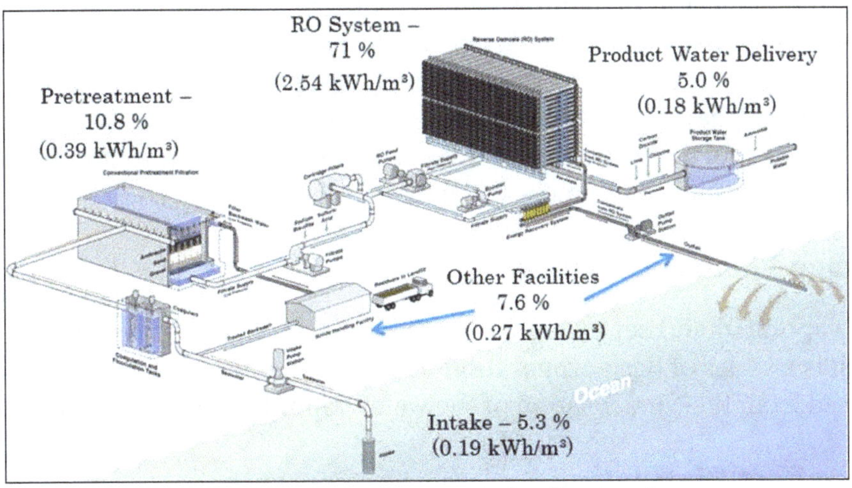

Plan of a typical reverse osmosis desalination plant.

Methods

Reverse osmosis desalination plant in Barcelona, Spain.

There are several methods. Each has advantages and disadvantages but all are useful.

The traditional process of desalination is distillation, i.e. boiling and re-condensation of seawater to leave salt and impurities behind.

Solar Distillation

Solar distillation mimics the natural water cycle, in which the sun heats the sea water enough for evaporation to occur. After evaporation, the water vapor is condensed onto a cool surface. There are two types of solar desalination. The former one is using photo voltaic cells which converts solar energy to electrical energy to power the desalination process. The later one utilises the solar energy in the heat form itself and is known as solar thermal powered desalination.

Vacuum Distillation

In vacuum distillation atmospheric pressure is reduced, thus lowering the temperature required to evaporate the water. Liquids boil when the vapor pressure equals the ambient pressure and vapor pressure increases with temperature. Effectively, liquids boil at a lower temperature, when the ambient atmospheric pressure is less than usual atmospheric pressure. Thus, because of the reduced pressure, low-temperature "waste" heat from electrical power generation or industrial processes can be employed.

Multi-stage Flash Distillation

Water is evaporated and separated from sea water through multi-stage flash distillation, which is a series of flash evaporations. Each subsequent flash process utilizes energy released from the condensation of the water vapor from the previous step.

Multiple-effect Distillation

Multiple-effect distillation (MED) works through a series of steps called "effects" Incoming water is sprayed onto pipes which are then heated to generate steam. The steam is then used to heat the next batch of incoming sea water. To increase efficiency, the steam used to heat the sea water can be taken from nearby power plants. Although this method is the most thermodynamically efficient among methods powered by heat, a few limitations exist such as a max temperature and max number of effects.

Vapor-compression Distillation

Vapor-compression evaporation involves using either a mechanical compressor or a jet stream to compress the vapor present above the liquid. The compressed vapor is then used to provide the heat needed for the evaporation of the rest of the sea water. Since this system only requires power, it is more cost effective if kept at a small scale.

Reverse Osmosis

The leading process for desalination in terms of installed capacity and yearly growth is reverse osmosis (RO). The RO membrane processes use semipermeable membranes and applied pressure (on the membrane feed side) to preferentially induce water permeation through the membrane while rejecting salts. Reverse osmosis plant membrane systems typically use less energy than thermal desalination processes. Energy cost in desalination processes varies considerably depending on water salinity, plant size and process type. At present the cost of seawater desalination, for example, is higher than traditional water sources, but it is expected that costs will continue to decrease with technology improvements that include, but are not limited to, improved efficiency, reduction in plants footprint, improvements to plant operation and optimization, more effective feed pretreatment, and lower cost energy sources.

Reverse osmosis uses a thin-film composite membrane, which comprises an ultra-thin, aromatic polyamide thin-film. This polyamide film gives the membrane its transport properties, whereas the remainder of the thin-film composite membrane provides mechanical support. The polyamide film is a dense, void-free polymer with a high surface area, allowing for its high water permeability.

The reverse osmosis process is not maintenance free. Various factors interfere with efficiency: ionic contamination (calcium, magnesium etc.); DOC; bacteria; viruses; colloids and insoluble particulates; biofouling and scaling. In extreme cases, the RO membranes are destroyed. To mitigate damage, various pretreatment stages are introduced. Anti-scaling inhibitors include acids and other agents such as the organic polymers polyacrylamide and polymaleic acid, phosphonates and polyphosphates. Inhibitors for fouling are biocides (as oxidants against bacteria and viruses), such as chlorine, ozone, sodium or calcium hypochlorite. At regular intervals, depending on the membrane contamination; fluctuating seawater conditions; or when prompted by monitoring processes, the membranes need to be cleaned, known as emergency or shock-flushing. Flushing is done with inhibitors in a fresh water solution and the system must go offline. This procedure is environmentally risky, since contaminated water is diverted into the ocean without treatment. Sensitive marine habitats can be irreversibly damaged.

Off-grid solar-powered desalination units use solar energy to fill a buffer tank on a hill with seawater. The reverse osmosis process receives its pressurized seawater feed in non-sunlight hours by gravity, resulting in sustainable drinking water production without the need for fossil fuels, an electricity grid or batteries. Nano-tubes are also used for the same function (i.e, Reverse Osmosis).

Freeze-thaw

Freeze-thaw desalination uses freezing to remove fresh water from salt water. Salt water is sprayed during freezing conditions into a pad where an ice-pile builds up. When seasonal conditions warm, naturally desalinated melt water is recovered. This technique relies on extended periods of natural sub-freezing conditions.

A different freeze-thaw method, not weather dependent and invented by Alexander Zarchin, freezes seawater in a vacuum. Under vacuum conditions the ice, desalinated, is melted and diverted for collection and the salt is collected.

Electrodialysis Membrane

Electrodialysis utilizes electric potential to move the salts through pairs of charged membranes, which trap salt in alternating channels.

Membrane Distillation

Membrane distillation uses a temperature difference across a membrane to evaporate vapor from a salty brine solution and condense pure condensate on the colder side.

Wave-powered Desalination

CETO is a wave power technology that desalinates seawater using submerged buoys. Wave-powered desalination plants began operating on Garden Island in Western Australia in 2013 and in Perth in 2015.

Considerations and Criticism

Energy Consumption

Energy consumption of seawater desalination has reached as low as 3 kWh/m³, including pre-filtering and ancillaries, similar to the energy consumption of other fresh water supplies transported over large distances, but much higher than local fresh water supplies that use 0.2 kWh/m³ or less.

A minimum energy consumption for seawater desalination of around 1 kWh/m³ has been determined, excluding prefiltering and intake/outfall pumping. Under 2 kWh/m³ has been achieved with reverse osmosis membrane technology, leaving limited scope for further energy reductions.

Supplying all US domestic water by desalination would increase domestic energy consumption by around 10%, about the amount of energy used by domestic refrigerators. Domestic consumption is a relatively small fraction of the total water usage.

Energy consumption of seawater desalination methods.				
Desalination Method >>	Multi-stage Flash MSF	Multi-Effect Distillation MED	Mechanical Vapor Compression MVC	Reverse Osmosis RO
Electrical energy (kWh/m³)	4–6	1.5–2.5	7–12	3–5.5

Thermal energy (kWh/m³)	50 110	60–110	None	None
Electrical equivalent of thermal energy (kWh/m³)	9.5–19.5	5–8.5	None	None
Total equivalent electrical energy (kWh/m³)	13.5–25.5	6.5–11	7–12	3–5.5

"Electrical equivalent" refers to the amount of electrical energy that could be generated using a given quantity of thermal energy and appropriate turbine generator. These calculations do not include the energy required to construct or refurbish items consumed in the process.

Cogeneration

Cogeneration is generating excess heat and electricity generation from a single process. Cogeneration can provide usable heat for desalination in an integrated, or "dual-purpose", facility where a power plant provides the energy for desalination. Alternatively, the facility's energy production may be dedicated to the production of potable water (a stand-alone facility), or excess energy may be produced and incorporated into the energy grid. Cogeneration takes various forms, and theoretically any form of energy production could be used. However, the majority of current and planned cogeneration desalination plants use either fossil fuels or nuclear power as their source of energy. Most plants are located in the Middle East or North Africa, which use their petroleum resources to offset limited water resources. The advantage of dual-purpose facilities is they can be more efficient in energy consumption, thus making desalination more viable.

The current trend in dual-purpose facilities is hybrid configurations, in which the permeate from reverse osmosis desalination is mixed with distillate from thermal desalination. Basically, two or more desalination processes are combined along with power production. Such facilities have been implemented in Saudi Arabia at Jeddah and Yanbu.

The Shevchenko BN-350, a former nuclear-heated desalination unit in Kazakhstan.

A typical Supercarrier in the US military is capable of using nuclear power to desalinate 1,500,000 L of water per day.

Economics

Costs of desalinating sea water (infrastructure, energy, and maintenance) are generally higher than fresh water from rivers or groundwater, water recycling, and water conservation, but alternatives are not always available. Desalination costs in 2013 ranged from US$0.45 to $1.00/cubic metre. More than half of the cost comes directly from energy cost, and since energy prices are very volatile, actual costs can vary substantially.

The cost of untreated fresh water in the developing world can reach US$5/cubic metre.

Average water consumption and cost of supply by sea water desalination at US$1 per cubic metre(±50%)		
Area	Consumption Litre/person/day	Desalinated Water Cost US$/person/day
USA	378	0.38
Europe	189	0.19
Africa	57	0.06
UN recommended minimum	49	0.05

Factors that determine the costs for desalination include capacity and type of facility, location, feed water, labor, energy, financing and concentrate disposal. Desalination stills control pressure, temperature and brine concentrations to optimize efficiency. Nuclear-powered desalination might be economical on a large scale.

While noting costs are falling, and generally positive about the technology for affluent areas in proximity to oceans, a 2004 study argued, "Desalinated water may be a solution for some water-stress regions, but not for places that are poor, deep in the interior of a continent, or at high elevation. Unfortunately, that includes some of the places with biggest water problems.", and, "Indeed, one needs to lift the water by 2000 m, or transport it over more than 1600 km to get transport costs equal to the desalination costs. Thus, it may be more economical to transport fresh water from somewhere else than to desalinate it. In places far from the sea, like New Delhi, or in high places, like Mexico City, transport costs could match desalination costs. Desalinated water is also expensive in places that are both somewhat far from the sea and somewhat high, such as Riyadh and Harare. By contrast in other locations transport costs are much less, such as Beijing, Bangkok, Zaragoza, Phoenix, and, of course, coastal cities like Tripoli." After desalination at Jubail, Saudi Arabia, water is pumped 320 km inland to Riyadh. For coastal cities, desalination is increasingly viewed as a competitive choice.

In 2014, the Israeli facilities of Hadera, Palmahim, Ashkelon, and Sorek were desalinizing water for less than US$0.40 per cubic meter. As of 2006, Singapore was desalinating water for US$0.49 per cubic meter. Perth began operating a reverse osmosis seawater desalination plant in 2006. A desalination plant now operates in Sydney, and the Wonthaggi desalination plant was under construction in Wonthaggi, Victoria.

The Perth desalination plant is powered partially by renewable energy from the Emu Downs Wind Farm. A wind farm at Bungendore in New South Wales was purpose-built to generate enough renewable energy to offset the Sydney plant's energy use, mitigating concerns about harmful greenhouse gas emissions.

In December 2007, the South Australian government announced it would build a seawater desalination plant for the city of Adelaide, Australia, located at Port Stanvac. The desalination plant was to be funded by raising water rates to achieve full cost recovery.

In November, Connecticut-based Poseidon Resources Corp. won a key regulatory approval to build the $300 million water-desalination plant in Carlsbad, north of San Diego. The facility would produce 190,000 cubic metres of drinking water per day, enough to supply about 100,000 homes. As of June 2012, the cost for the desalinated water had risen to $2,329 per acre-foot. Each $1,000 per acre-foot works out to $3.06 for 1,000 gallons, or $.81 per cubic meter.

Poseidon Resources made an unsuccessful attempt to construct a desalination plant in Tampa Bay, FL, in 2001. The board of directors of Tampa Bay Water was forced to buy the plant from Poseidon in 2001 to prevent a third failure of the project. Tampa Bay Water faced five years of engineering problems and operation at 20% capacity to protect marine life. The facility reached capacity only in 2007.

In 2008, a Energy Recovery Inc. was desalinating water for $0.46 per cubic meter.

Environmental

Factors that determine the costs for desalination include capacity and type of facility, location, feed water, labor, energy, financing and concentrate disposal.

Intake

In the United States, cooling water intake structures are regulated by the Environmental Protection Agency (EPA). These structures can have the same impacts to the environment as desalination facility intakes. According to EPA, water intake structures cause adverse environmental impact by sucking fish and shellfish or their eggs into an industrial system. There, the organisms may be killed or injured by heat, physical stress, or chemicals. Larger organisms may be killed or injured when they become trapped against screens at the front of an intake structure. Alternative intake types that mitigate these impacts include beach wells, but they require more energy and higher costs.

The Kwinana Desalination Plant opened in Perth in 2007. Water there and at Queensland's Gold Coast Desalination Plant and Sydney's Kurnell Desalination Plant is withdrawn at 0.1 m/s (0.33 ft/s), which is slow enough to let fish escape. The plant provides nearly 140,000 m^3 (4,900,000 cu ft) of clean water per day.

Desalination processes produce large quantities of brine, possibly at above ambient temperature, and contain residues of pretreatment and cleaning chemicals, their reaction byproducts and heavy metals due to corrosion. Chemical pretreatment and cleaning are a necessity in most desalination plants, which typically includes prevention of biofouling, scaling, foaming and corrosion in thermal plants, and of biofouling, suspended solids and scale deposits in membrane plants.

To limit the environmental impact of returning the brine to the ocean, it can be diluted with another stream of water entering the ocean, such as the outfall of a wastewater treatment or power plant. With medium to large power plant and desalination plants, the power plant's cooling water flow is likely to be several times larger than that of the desalination plant, reducing the salinity of the combination. Another method to dilute the brine is to mix it via a diffuser in a mixing zone. For example, once a pipeline containing the brine reaches the sea floor, it can split into many branches, each releasing brine gradually through small holes along its length. Mixing can be combined with power plant or wastewater plant dilution.

Brine is denser than seawater and therefore sinks to the ocean bottom and can damage the ecosystem. Careful reintroduction can minimize this problem.

Alternatives to Desalination

Increased water conservation and efficiency remain the most cost-effective approaches in areas with a large potential to improve the efficiency of water use practices. Wastewater reclamation provides multiple benefits over desalination. Urban runoff and storm water capture also provide benefits in treating, restoring and recharging groundwater.

A proposed alternative to desalination in the American Southwest is the commercial importation of bulk water from water-rich areas either by oil tankers converted to water carriers, or pipelines. The idea is politically unpopular in Canada, where governments imposed trade barriers to bulk water exports as a result of a North American Free Trade Agreement (NAFTA) claim.

Public Health Concerns

Desalination removes iodine from water and could increase the risk of iodine deficiency disorders. Israeli researchers claimed a possible link between seawater desalination and iodine deficiency, finding deficits among euthyroid adults exposed to iodine-poor water concurrently with an increasing proportion of their area's drinking water from seawater reverse osmosis (SWRO). They later found probable iodine deficiency disorders in a population reliant on desalinated seawater. A possible link of heavy desalinated water use and national iodine deficiency was suggested by Israeli researchers. They found a high burden of iodine deficiency in the general population of Israel: 62% of school-age children and 85% of pregnant women fall below the WHO's adequacy range. They also pointed out the national reliance on iodine-depleted desalinated water, the absence of a universal salt iodization program and reports of increased use of thyroid medication in Israel as a possible reasons that the population's iodine intake is low. In the year that the survey was conducted, the amount of water produced from the desalination plants constitutes about 50% of the quantity of fresh water supplied for all needs and about 80% of the water supplied for domestic and industrial needs in Israel.

Other Issues

Due to the nature of the process, there is a need to place the plants on approximately 25 acres of land on or near the shoreline. In the case a plant is built inland, pipes have to be laid into the ground to allow for easy intake and outtake. However, once the pipes are laid into the ground, they have a possibility of leaking into and contaminating nearby aquifers. Aside from environmental risks, the noise generated by certain types of desalination plants can be loud.

Public Opinion

Despite the issues associated with desalination processes, public support for its development can be very high. One survey of a Southern California community saw 71.9% of all respondents being in support of desalination plant development in their community. In many cases, high freshwater scarcity corresponds to higher public support for desalination development whereas areas with low water scarcity tend to have less public support for its development.

Experimental Techniques

Other desalination techniques include:

Waste Heat

Thermally-driven desalination technologies are frequently suggested for use with low-temperature waste heat sources, as the low temperatures are not useful for many industrial processes, but ideal for the lower temperatures found in desalination. In fact, such pairing with waste heat can even improve electrical process: Diesel generators commonly provide electricity in remote areas. About 40%–50% of the energy output is low-grade heat that leaves the engine via the exhaust. Connecting a thermal desalination technology such as membrane distillation system to the diesel engine exhaust repurposes this low-grade heat for desalination. The system actively cools the diesel generator, improving its efficiency and increasing its electricity output. This results in an energy-neutral desalination solution. An example plant was commissioned by Dutch company Aquaver in March 2014 for Gulhi, Maldives.

Low-temperature Thermal

Originally stemming from ocean thermal energy conversion research, low-temperature thermal desalination (LTTD) takes advantage of water boiling at low pressure, even at ambient temperature. The system uses pumps to create a low-pressure, low-temperature environment in which water boils at a temperature gradient of 8–10 °C (46–50 °F) between two volumes of water. Cool ocean water is supplied from depths of up to 600 m (2,000 ft). This water is pumped through coils to condense the water vapor. The resulting condensate is purified water. LTTD may take advantage of the temperature gradient available at power plants, where large quantities of warm wastewater are discharged from the plant, reducing the energy input needed to create a temperature gradient.

Experiments were conducted in the US and Japan to test the approach. In Japan, a spray-flash evaporation system was tested by Saga University. In Hawaii, the National Energy Laboratory tested an open-cycle OTEC plant with fresh water and power production using a temperature difference of 20 C° between surface water and water at a depth of around 500 m (1,600 ft). LTTD was studied by India's National Institute of Ocean Technology (NIOT) in 2004. Their first LTTD plant opened in 2005 at Kavaratti in the Lakshadweep islands. The plant's capacity is 100,000 L (22,000 imp gal; 26,000 US gal)/day, at a capital cost of INR 50 million (€922,000). The plant uses deep water at a temperature of 10 to 12 °C (50 to 54 °F). In 2007, NIOT opened an experimental, floating LTTD plant off the coast of Chennai, with a capacity of 1,000,000 L (220,000 imp gal; 260,000 US gal)/day. A smaller plant was established in 2009 at the North Chennai Thermal Power Station to prove the LTTD application where power plant cooling water is available.

Thermionic Process

In October 2009, Saltworks Technologies announced a process that uses solar or other thermal heat to drive an ionic current that removes all sodium and chlorine ions from the water using ion-exchange membranes.

Evaporation and Condensation for Crops

The Seawater greenhouse uses natural evaporation and condensation processes inside a greenhouse powered by solar energy to grow crops in arid coastal land.

Other Approaches

Adsorption-based desalination (AD) relies on the moisture absorption properties of certain materials such as Silica Gel.

Forward Osmosis

One process was commercialized by Modern Water PLC using forward osmosis, with a number of plants reported to be in operation.

Hydrogel based Desalination

The idea of the method is in the fact that when the hydrogel is put into contact with aqueous salt solution, it swells absorbing a solution with the ion composition different from the original one. This solution can be easily squeezed out from the gel by means of sieve or microfiltration membrane. The compression of the gel in closed system lead to change in salt concentration, whereas the compression in open system, while the gel is exchanging ions with bulk, lead to the change in the number of ions. The consequence of the compression and swelling in open and closed system conditions mimics the reverse Carnot Cycle of refrigerator machine. The only difference is that instead of heat this cycle transfers salt ions from the bulk of low salinity to a bulk of high salinity. Similarly to the Carnot cycle this cycle is fully reversible, so can in principle work with an ideal thermodynamic efficiency. Because the method is free from the use of osmotic membranes it can compete with reverse osmosis method. In addition, unlike the reverse osmosis, the approach is not sensitive to the quality of feed water and its seasonal changes, and allows the production of water of any desired concentration.

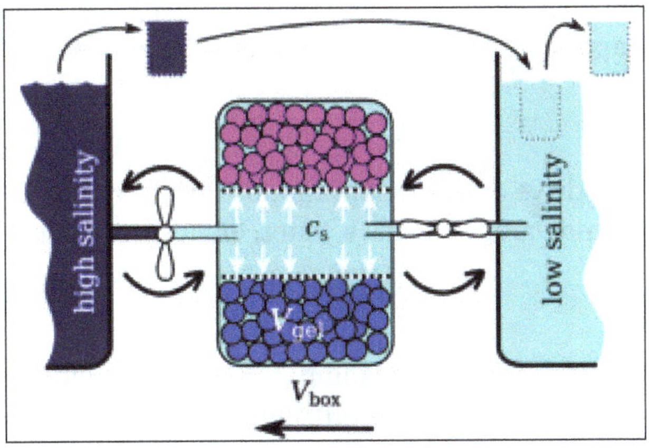

Desalination flow of ions.

Scheme of the desalination machine: the desalination box of volume V_{box} contains a gel of volume V_{gel} which is separated by a sieve from the outer solution volume $V_{out} = V_{box} - V_{gel}$. The box is connected to two big tanks with high and low salinity by two taps which can be opened and closed as desired. The chain of buckets expresses the fresh water consumption followed by refilling the low-salinity reservoir by salt water.

Small-scale Solar

The United States, France and the United Arab Emirates are working to develop practical solar desalination. AquaDania's WaterStillar has been installed at Dahab, Egypt, and in Playa del Carmen, Mexico. In this approach, a solar thermal collector measuring two square metres can distill from 40 to 60 litres per day from any local water source – five times more than conventional stills. It eliminates the need for plastic PET bottles or energy-consuming water transport. In Central California, a startup company WaterFX is developing a solar-powered method of desalination that can enable the use of local water, including runoff water that can be treated and used again. Salty groundwater in the region would be treated to become freshwater, and in areas near the ocean, seawater could be treated.

Passarell

The Passarell process uses reduced atmospheric pressure rather than heat to drive evaporative desalination. The pure water vapor generated by distillation is then compressed and condensed using an advanced compressor. The compression process improves distillation efficiency by creating the reduced pressure in the evaporation chamber. The compressor centrifuges the pure water vapor after it is drawn through a demister (removing residual impurities) causing it to compress against tubes in the collection chamber. The compression of the vapor increases its temperature. The heat is transferred to the input water falling in the tubes, vaporizing the water in the tubes. Water vapor condenses on the outside of the tubes as product water. By combining several physical processes, Passarell enables most of the system's energy to be recycled through its evaporation, demisting, vapor compression, condensation, and water movement processes.

Geothermal

Geothermal energy can drive desalination. In most locations, geothermal desalination beats using scarce groundwater or surface water, environmentally and economically.

Nanotechnology

Nanotube membranes of higher permeability than current generation of membranes may lead to eventual reduction in the footprint of RO desalination plants. It has also been suggested that the use of such membranes will lead to reduction in the energy needed for desalination.

Hermetic, sulphonated nano-composite membranes have shown to be capable of removing a various contaminants to the parts per billion level. s, have little or no susceptibility to high salt concentration levels.

Biomimesis

Biomimetic membranes are another approach.

Electrochemical

In 2008, Siemens Water Technologies announced technology that applied electric fields to desalinate one cubic meter of water while using only a purported 1.5 kWh of energy. If accurate, this process would consume one-half the energy of other processes. As of 2012 a demonstration plant was operating in Singapore. Researchers at the University of Texas at Austin and the University of Marburg are developing more efficient methods of electrochemically mediated seawater desalination.

Electrokinetic Shocks

A process employing electrokinetic shocks waves can be used to accomplish membraneless desalination at ambient temperature and pressure. In this process, anions and cations in salt water are exchanged for carbonate anions and calcium cations, respectively using electrokinetic shockwaves. Calcium and carbonate ions react to form calcium carbonate, which precipitates, leaving fresh water. The theoretical energy efficiency of this method is on par with electrodialysis and reverse osmosis.

Temperature Swing Solvent Extraction

A group of researchers from Columbia University have developed a process designed to purify industrial hypersaline brines. They call it Temperature Swing Solvent Extraction (TSSE). TSSE can desalinate extremely salty brine up to seven times as salty as the ocean. For comparison, the current methods can only handle brine twice as salty. Unlike other approaches that use reverse osmosis or distillation, TSSE has found success with using a solvent. Solvent extraction is commonly used during chemical engineering. The TSSE solvent isn't dependent on the evaporation of water, meaning it doesn't need high temperatures to work. It can be activated by low-grade heat (less than 70 degrees celsius) that is easy to attain, sometimes to the point of it being natural. In a study, TSSE removed up to 98.4 percent of the salt in brine.

Facilities

Estimates vary widely between 15,000–20,000 desalination plants producing more than 20,000 m^3/day. Micro desalination plants operate near almost every natural gas or fracking facility found in the United States.

In Nature

Mangrove leaf with salt crystals.

Evaporation of water over the oceans in the water cycle is a natural desalination process.

The formation of sea ice produces ice with little salt, much lower than in seawater.

Seabirds distill seawater using countercurrent exchange in a gland with a rete mirabile. The gland secretes highly concentrated brine stored near the nostrils above the beak. The bird then "sneezes" the brine out. As freshwater is not usually available in their environments, some seabirds, such as pelicans, petrels, albatrosses, gulls and terns, possess this gland, which allows them to drink the salty water from their environments while they are far from land.

Mangrove trees grow in seawater; they secrete salt by trapping it in parts of the root, which are then eaten by animals (usually crabs). Additional salt is removed by storing it in leaves that fall off. Some types of mangroves have glands on their leaves, which work in a similar way to the seabird desalination gland. Salt is extracted to the leaf exterior as small crystals, which then fall off the leaf.

Willow trees and reeds absorb salt and other contaminants, effectively desalinating the water. This is used in artificial constructed wetlands, for treating sewage.

HOUSEHOLD APPLICATIONS

Low-flush Toilet

A low-flush toilet (or low-flow toilet or high-efficiency toilet) is a flush toilet that uses significantly less water than a full-flush toilet. Low-flush toilets use 4.8 litres (1.3 US gal; 1.1 imp gal) or less per flush, as opposed to 6 litres (1.6 US gal; 1.3 imp gal) or more. They came into use in the United States in the 1990s, in response to water conservation concerns. Low-flush toilets include single-flush models and dual-flush toilets, which typically use 1.6 USgpf for the full flush and 1.28 US for a reduced flush.

Water Savings

The US Environmental Protection Agency's WaterSense program provides certification that toilets meet the goal of using less than 1.6 US gallons per flush. Units that meet or exceed this standard can carry the WaterSense sticker. The EPA estimates that the average US home will save US$90 per year, and $2,000 over the lifetime of the toilets. Dry toilets can lead to even more water savings in private homes as they use no water for flushing.

Problems

The early low-flush toilets in the U.S. often had a poor design that required more than one flush to rid the bowl of solid waste, resulting in more water used than a single flush of a standard toilet. In response, U.S. Congressman Joe Knollenberg from Michigan tried to get Congress to repeal the law[clarification needed] but was unsuccessful.

In 2011, the San Francisco Chronicle reported that, while low-flush toilets are estimated to have saved the city of San Francisco 20 million gallons of water per year, the reduction in water volume caused waste sludge to back up in the city sewer pipes, designed expecting a higher ratio of water to solids. The city was attempting to solve this by adding chlorine bleach to the pipes, a proposal that raised environmental objections. In house drain system design, smaller-diameter drain pipes are being used to improve flow by forcing waste to run higher in the pipe and therefore have less tendency to settle along the pipe.

Dry Toilet

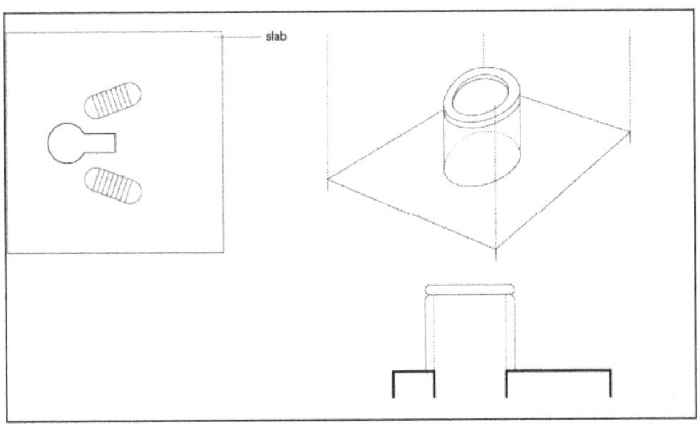

Schematic of a dry toilet: Left a squat toilet, right a pedestal type toilet.

A dry toilet (or non-flush toilet, no flush toilet or toilet without a flush) is a toilet that operates without flush water, unlike a flush toilet. The dry toilet may have a raised pedestal on which the user can sit, or a squat pan over which the user squats in the case of a squat toilet. In both cases, the excreta (both urine and feces) falls through a drop hole.

A dry toilet can be any of the following types of toilets: a composting toilet, urine-diverting dry toilet, arborloo, container-based toilet, bucket toilet, simple pit latrine (but not those that operate on a "pour flush" basis), incinerating toilets, or freezing toilets.

The urine and feces can either become mixed at the point of dropping or stay separated, which is called urine diversion.

Types

There are several types of toilets which are referred to as "dry toilets". All of them work without flush water and without a connection to a sewer system or septic tank:

- Composting toilet (in most cases without urine diversion but can also be with urine diversion).

- Urine-diverting dry toilet (UDDT): With urine diversion, as the name implies.
- Arborloo: Which is similar to a pit latrine but has a much shallower pit and is designed for making compost in the pit.
- Container-based toilet: Where human excreta is collected in sealable, removable containers (also called cartridges) that are transported to treatment facilities (a sub-type are "packaging toilets" where feces from each defecation event is placed in its own package).
- Bucket toilet: A very basic type of dry toilet consisting only of a bucket which could be upgraded with urine diversion and the use of covering material.
- Simple pit latrines: Those pit latrines that do not have a water seal (pour flush pit latrines have a water seal and are therefore not dry toilets).
- Incinerating toilets, freezing toilets: These are toilets with more complicated technology and higher costs.

Other types of dry toilets are under development at universities, for example since 2012 funded by the Bill and Melinda Gates Foundation. Such toilets are meant to operate off-the-grid without connections to water, sewer, or electrical lines.

References

- A-guide-to-water-conservation, pages: theberkey.com, Retrieved 7 July, 2020
- Black, Maggie (2009). The Atlas of Water. Berkeley and Los Angeles, California: University of California Press. P. 62. ISBN 9780520259348
- Outstanding-water-conservation-methods: biologydiscussion.com, Retrieved 8 August, 2020
- Callard, Sarah (2007-09-29). "Is it worth it? Installing dual flush". Daily Telegraph (UK). Retrieved 2019-05-31
- Types-of-waterless-urinal-technologies-39aeeb2e5558: ekameco.com, Retrieved 9 January, 2020
- Hamed, O. A. (2005). "Overview of hybrid desalination systems – current status and future prospects". Desalination. 186 (1–3): 207. CiteSeerX 10.1.1.514.4201. doi:10.1016/j.desal.2005.03.095
- The-pros-and-cons-of-waterless-urinals: waterless.com, Retrieved 10 February, 2020
- Indira, khan. "Quooker kranen kopen doet u hier". Kokendwaterkranen.com (in Dutch). Retrieved 20 December 2020
- Drip-irrigation-system, farm-machinary, agri-inputs, agriculture: vikaspedia.in, Retrieved 11 March, 2020
- Shaw, R. (2014). A Collection of Contemporary Toilet Designs. EOOS and WEDC, Loughborough University, UK. P. 40. ISBN 978 1 84380 155 9
- Surface-irrigation-methods: aboutcivil.org, Retrieved 12 April, 2020
- Biggs, John (7 December 2015). "Hamwells e-Shower Connects Your Shower To The Internet Of Things (And Bathing)". Techcrunch. Retrieved 15 March 2017

4

Water Resource Management

The activity of planning, developing, distributing and managing water resources to ensure their optimum use is referred to as water management. It is primarily concerned with water resource management, irrigation management, integrated urban water management and storm water management. All these diverse aspects of water management have been carefully analyzed in this chapter.

Water Management is important since it helps determine future irrigation expectations. Water management is the management of water resources under set policies and regulations. Water, once an abundant natural resource, is becoming a more valuable commodity due to droughts and overuse. Here are links to articles that address water management subjects such as the optimization of water usage. Water resource management is the activity of planning, developing, distributing and managing the optimum use of water resources.

It is a sub-set of water cycle management. Ideally, water resource management planning has regard to all the competing demands for water and seeks to allocate water on an equitable basis to satisfy all uses and demands. As with other resource management, this is rarely possible in practice. Water is an essential resource for all life on the planet. Of the water resources on Earth only three percent of it is fresh and two-thirds of the freshwater is locked up in ice caps and glaciers. Of the remaining one percent, a fifth is in remote, inaccessible areas and much seasonal rainfall in monsoonal deluges and floods cannot easily be used. At present only about 0.08 percent of all the world's fresh water is exploited by mankind in ever increasing demand for sanitation, drinking, manufacturing, leisure and agriculture.

Water resource management includes consideration of all of the above disciplines of hydrology. Water supplies are allocated and diverted to a range of agricultural, municipal, industrial, hydroelectrical, and ecological needs. Some of these water uses are consumptive, removing water from the system (e.g., crop irrigation). Other types of water use return the water to a river, lake, or to the ground, but the water often requires treatment to restore it to a natural state; sometimes this is not possible (e.g., industrial tailings ponds).

The balancing act involved in water management includes a broad range of stakeholders and includes water policy and legal experts. Hydrologists have essential input to these complex and sometimes confrontational deliberations and negotiations. They also play a central role in applied hydrology – engineering of major waterworks to manage water. Water distribution systems have been a hallmark of civilization since Babylon, and the modern stamp on this includes major hydroelectric dams and reservoirs, urban waterworks, and water treatment facilities.

These and other tools help governments to manage water resources in a way that serves societal and ecological needs. However, water resource management is one of the world's greatest challenges due to competition for limited resources, regional disparities in water supply and affluence, mounting global water demand, aquifer depletion, and pollution- and climate-change induced water stress. Integrated sustainable water resource management is an area requiring innovation, progress, and international cooperation in the coming decades.

A central aspect of water resource management is the concept of setting sustainable resource limits and then allocating water (licensing) within these limits. A key approach for managing diffuse causes of hydrological alteration is to bring these water uses within the resource limit and license these activities or their impacts. It is far easier to make the resource limits comprehensive from the outset rather than retrospectively regulate water use.

A key challenge in regulating diffuse activities is monitoring and compliance across large spatial scales and potentially a large number of entities. One approach that may be useful is the concept of responsive regulation, an approach developed by Ayres and Braithwaite. Compliance with regulations is most likely to occur when a regulator uses an approach that Ayres and Braithwaite refer to as a regulatory pyramid. At the base of the pyramid are dialogue-based approaches for securing compliance such as persuasion and education. Ayres and Braithwaite suggest that these should be the most commonly used techniques. The higher levels of the pyramid consist of more demanding and punitive interventions such as formal warnings, penalties, and license revocation. Braithwaite argues that regulators should always start at the base of the pyramid before reluctantly escalating to more severe approaches in cases when dialogue fails, and only employing more punitive approaches when the more modest forms of punishment have failed. This technique is used by Melbourne Water, an Australian water agency, to enforce farm dam laws. As with many environmental policies, public education to change perceptions may contribute toward a successful.

Economic implications of water resources management are important when considering the decision-making process related to environment and health risks in developing countries. In most of these areas, the diseases associated with poor water and sanitation have an extremely relevant public health impact. Cost-benefit analysis (CBA)

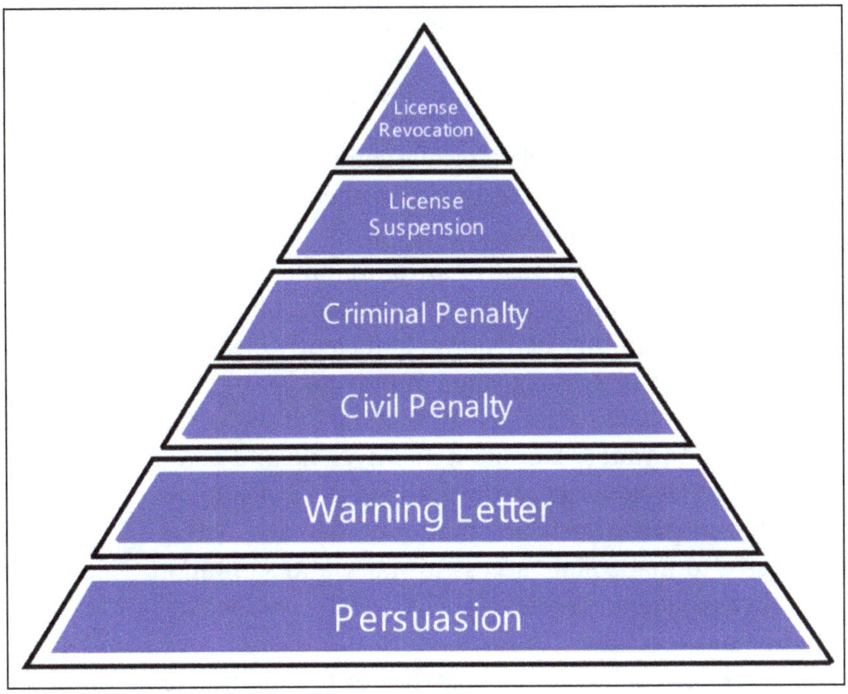

An example of a regulatory pyramid. The size of each layer suggests the appropriate proportion of activity at each level. The content of the pyramid will vary depending on the context of the particular regulation.

is the form of economic evaluation most useful for resource allocation investigation between government-financed activities and within the productive sectors. The aim of this topic is to outline the main costs and benefits associated with investments destined to improve Water Supply and Sanitation Services (WSS). It is important to underline that when considering this category of interventions, the objective is to measure their overall welfare impact, even if it is not possible to disentangle completely the financial impact of all benefits.

The annual costs of improving access to safe water supply and adequate sanitation services vary depending mainly on the increase in the additional number of people being served. The cost assessment of WSS interventions is not well documented. The main sources of cost data are the Global Water Supply and Sanitation Assessment Reports of 2000 and 2017. In these reports, however, the interventions that are considered do usually not concern specific countries. This is particularly true for the countries in the MENA region, since household surveys and ministry statistics are not regularly performed and updated. Existing cost-benefit analyses calculate costs as the sum of all resources required to put in place and maintain the infrastructures. These costs include investment costs (planning, hardware construction, and house alteration) and recurrent costs (operation and maintenance of hardware, replacement of parts, emptying of septic tank, latrines, education, monitoring, regulation, and control of water supply).

Recurrent costs are usually computed as a percentage of the annual investment cost. An estimate of the total cost per year in the MENA region is around 6 billion of US dollar that means around 13 US dollar of annual cost per capita.

On the other hand, investment in WSS typically generates several economic, environmental, and social benefits. Improving water quality and extending water access reduce health risks and frees-up time for education and other productive activities, it also increases the productivity of the labor force, because of lower malnutrition. Safe disposal of wastewaters helps to improve the quality of surface waters with benefits for the environment, as well as for economic sectors that depend on sufficiently clean water as a resource. The most significant impact of improving water supply and sanitation services is on health. From a global perspective, approximately 10% of the burden of disease could be prevented with improvements to water, sanitation and hygiene and better water resource management. Focusing on developing countries, the burden of water-related diseases counts disproportionately and particularly on children under five, with 30% of deaths of these children attributable to inadequate access to water and sanitation. This corresponds to 61 million disability-adjusted life-years (DALYs) lost, considering burden of disease due to both morbidity and mortality. Unfortunately, there is no specific information on all waterborne diseases per each country: diarrhea is the main disease burden from poor water and sanitation for which there is data for all regions on incidence rates and deaths. Studies on benefits from WSS interventions consider only diarrhea related data. Therefore, their results underestimate the effective positive benefit on health. To evaluate the benefits, two main outcomes that are associated with waterborne diseases are considered: the reduction in incidence rates and the reduction in the number of deaths.

Once these outcomes are assessed, it is possible to evaluate how DALYs change in correspondence to their reduction. The cost savings associated with reduction in incidence rates are calculated by multiplying the health service unit cost by the number of cases averted. Nonhealth sector direct costs are mainly those that fall on the patient, such as transport costs to health services, other visit expenses and the opportunity costs of time. In terms of deaths avoided the expected number is predicted from the incidence rate multiplied by the case fatality rate and the economic impact is evaluated in terms of lack of productivity. The most reliable estimate of direct return to WSS investment in terms of health is up to 8 dollars every dollar invested: considering this, investments to improve drinking water, sanitation, hygiene, and water resource management systems make strong economic sense. In addition to the value of saved human lives and health-care savings, there are observable gains also in term of economic productivity. This benefit is somehow related to health but not directly. Health improvement affects productivity in two ways: lower morbidity and fewer deaths. Clustering the incidence of waterborne diseases per age, it would be possible to estimate the gain in productivity due to WSS interventions, multiplying for those of working age, the number of work days gained (usually assumed to be 2) per the individual economic value of time. To consider the gain due to fewer deaths the convention is to predict the economic costs of premature mortality, multiplying the number of productive years left and an estimate of the income that would be earned, and discounting the income to the present time period. But gains in productivity depend not only on health effects. A fundamental productive sector in that area is agriculture and irrigation is essential to increase crop productivity and quality. Irrigation provides farmers with direct benefits through higher and more stable incomes and a higher value of irrigated land. Communities benefit from better salaries, less expensive food, a healthier and diversified diet together with beneficial health effects due to increased water availability. Increasing productivity is the key activity to reduce poverty in many developing countries, in the MENA region as much as elsewhere, and investing in WSS is one of the measures through which this objective can most effectively be reached.

The economic benefit of investing in drinking-water and sanitation comes also in the form of time savings. Time savings occur due to, for example, the relocation of a well or borehole nearer populations, piped water supply to households, and closer access to latrines and less waiting time for public latrines. These time savings give either increased production or more leisure time: for example, since women are no longer forced to fetch water from far away, they have time to devote themselves to the horticultural crop, increasing income and improving their own diet and the one of their families. For children of school age, the impact is that a waterborne illness prevents them from going to school, thus interrupting their education. Given the recognized importance of proper schooling for future productivity as well as the overall welfare of society, it is important to take into account the social and economic implications of children missing school due to ill health. We can conclude that investments in getting easily access to water and in improving drinking-water quality create benefits of two types: those which can easily be evaluated in monetary terms through a cost-benefit analysis and those for which it is difficult to investigate the financial impact. Even if previous analyses suggest the financial efficiency in investing in WSS, this becomes stringer when adding social and environmental benefits. To improve the decision-making process that allows to make investments in water management it is important to coordinate activities between countries that share water resources.

Water resource management is of paramount importance in agricultural development. Sustained socioeconomic development in countries with irrigated agriculture could be limited by water availability and deterioration in water quality. Consequently, any effort made to improve water use in irrigation practices will have a marked effect on sustainable agriculture and on conservation of soil and water resources. Problems associated with irrigated agriculture have been attributed to poor water management due to lack of knowledge of soil–plant–atmosphere relations, inadequate form of water application, hydrochemical relations triggered by irrigation, and soil's microbiological surroundings.

Optimal design and selection of irrigation systems permit high efficiencies and uniform distribution of the applied water. Advanced planning and management of water is needed to ensure a sustainable agricultural process with optimal crop production, considering physical, environmental, financial, and technological restrictions. In addition, drainage techniques have also been developed, allowing removal of excess water from the soil profile. However, there is still a need to optimize water resource use and management to maintain sustainable irrigated agriculture considering water demand, water quality, economic evaluation of water, and decision support systems.

INTEGRATED URBAN WATER MANAGEMENT

Integrated urban water management (IUWM) is a philosophy of varying definitions and interpretations. According to the authors of the book entitled, "Integrated Urban Water Management: Humid Tropics", IUWM is described as the practice of managing freshwater, wastewater, and storm water as components of a basin-wide management plan. It builds on existing water supply and sanitation considerations within an urban settlement by incorporating urban water management within the scope of the entire river basin. One of the early champions of IUWM, SWITCH is a research program funded by the European Union and seeks to shift urban water management away from ad hoc solutions to a more integrated approach. IUWM within an urban water system can also be conducted by performance assessment of any new intervention strategies by developing a holistic approach which encompasses various system elements and criteria including sustainability type ones in which integration of water system components including water supply, waste water and storm water subsystems would be advantageous. Simulation of metabolism type flows in urban water system can also be useful for analysing processes in urban water cycle of IUWM.

IUWM is commonly seen as a strategy for achieving the goals of Water Sensitive Urban Design. IUWM seeks to change the impact of urban development on the natural water cycle, based on the premise that by managing the urban water cycle as a whole; a more efficient use of resources can be achieved providing not only economic benefits but also improved social and environmental outcomes. One approach is to establish an inner, urban, water cycle loop through the implementation of reuse strategies. Developing this urban water cycle loop requires an understanding both of the natural, pre-development, water balance and the post-development water balance. Accounting for flows in the pre- and post-development systems is an important step toward limiting urban impacts on the natural water cycle.

Components

Activities under the IUWM include the following:

- Improve water supply and consumption efficiency.
- Upgrade drinking water quality and wastewater treatment.

- Increase economic efficiency of services to sustain operations and investments for water, wastewater, and stormwater management.
- Utilize alternative water sources, including rainwater, and reclaimed and treated water.
- Engage communities to reflect their needs and knowledge for water management.
- Establish and implement policies and strategies to facilitate the above activities.
- Support capacity development of personnel and institutions that are engaged in IUWM.

According to Australia's Commonwealth Scientific and Industrial Research Organisation (CSIRO), IUWM requires the management of the urban water cycle in coordination with the hydrological water cycle which are significantly altered by urban landscapes and its correlation to increasing demand. Under natural conditions the water inputs at any point in the system are precipitation and overland flows; while the outputs are via surface flows, evapo-transpiration and groundwater recharge. The large volumes of piped water introduced with the change to an urban setting and the introduction of vast impervious areas strongly impact the water balance, increasing in-flows and dramatically altering the out-flow components.

Approaches

- The Agenda 21 (UN Department for Sustainable Development, 1992) has worked out the Dublin Principles for Integrated water resources management in more detail for urban areas. One of the objectives of Agenda 21 is to develop environmentally sound management of water resources for urban use.
- The Bellagio Statement formulated by the Environmental Sanitation Working Group of the Water Supply and Sanitation Collaborative Council in 2000 include principals such as: Human dignity, quality of life, environmental security, an open stakeholder process, and many others.
- The UNEP 3 Step Strategic Approach developed in 2005 is based on the application of the "Cleaner Production approach" that has been successful in the industrial sector. The three steps are: Prevention, Treatment for reuse, and Planned discharge with stimulation of self-purification capacity.
- UNESCO's Institute for Water Education seeks to build on the progress made by the Bellagio Statement and UNEP's 3-step approach by developing the *SWITCH* approach to IUWM. Components include: the addition of a sustainability assessment, new methods of planning urban water systems, and modifications to planning and strategy development.

An example of IUWM is the Catskill/ Delaware water system that provides 1.4 billion US gallons (5,300,000 m^3) of water per day, including to all of New York City. The IUWM process included an extensive stakeholder engagement process, whereby the needs of all parties were included into the final management plan. A partnership was created between New York City, the agricultural community, and the federal government. The case has become a model for successful IUWM.

Grey Water Systems

Grey water (also written; greywater, gray water, or graywater) is water used with appliances that do not involve or encounter human waste. It gets its name relative to black water which is heavily contaminated with human waste. Different resources suggest what equipment produce grey or black water. However, it is most commonly accepted that bathtubs, showers, washbasins, washing machines, and laundry tubs produce grey water, whereas toilets, sinks, and dishwashers are classified as black water sources.

In 1989 Santa Barbara became the first district in the United States to legalize the recycling and reuse of grey water. Since then, grey water has become a part of integrated urban water management. It addresses the practice of managing wastewater at the residential scale. The premise of grey water reuse is the concept that the average household uses most of its water indoors (roughly 60%), meaning much of that water can be reused to provide for the water required to support irrigation. Additionally, most domestic appliances automatically collect grey water in order for it to be disposed.

There are three types of grey water systems each of which has different requirements, codes, and sizing specifications. However, they share standards to meet health and safety regulations:

The Arizona R18-9-711. Type 1 Reclaimed water General Permit for Gray Water, stands as the model for grey water system regulation:

- Human contact with gray water and soil irrigated by gray water is avoided;
- Gray water originating from the residence is used and contained within the property boundary for household gardening, composting, lawn watering, or landscape irrigation;
- Surface application of gray water is not used for irrigation of food plants, except for citrus and nut trees;
- The gray water does not contain hazardous chemicals derived from activities such as cleaning car parts, washing greasy or oily rags, or disposing of waste solutions from home photo labs or similar hobbyist or home occupational activities;
- The application of gray water is managed to minimize standing water on the surface;

- The gray water system is constructed so that if blockage, plugging, or backup of the system occurs, gray water can be directed into the sewage collection system or onsite wastewater treatment and disposal system, as applicable. The gray water system may include a means of filtration to reduce plugging and extend system lifetime;
- Any gray water storage tank is covered to restrict access and to eliminate habitat for mosquitoes or other vectors;
- The gray water system is sited outside of a floodway;
- The gray water system is operated to maintain a minimum vertical separation distance of at least five feet from the point of gray water application to the top of the seasonally high groundwater table;
- For residences using an onsite wastewater treatment facility for black water treatment and disposal, the use of a gray water system does not change the design, capacity, or reserve area requirements for the onsite wastewater treatment facility at the residence, and ensures that the facility can handle the combined black water and gray water flow if the gray water system fails or is not fully used;
- Any pressure piping used in a gray water system that may be susceptible to cross connection with a potable water system clearly indicates that the piping does not carry potable water;
- Gray water applied by surface irrigation does not contain water used to wash diapers or similarly soiled or infectious garments unless the gray water is disinfected before irrigation;
- Surface irrigation by gray water is only by flood or drip irrigation.

A complex grey water system provides for a development with a substantial discharge (greater than 250 gallons) per day. It requires a written construction permit submitted to the enforcing agency.

A simple grey water system is sized to serve a one or two family home with a medium level water discharge (maximum 250 gallons) per day. It too, requires a written construction permit submitted to the Enforcing Agency.

A clothes washer grey water system is sized to recycle the grey water of a one or two family home using the reclaimed water of a washing machine (produces 15 gallons per person per day). It relies on either the pump from the washing machine or gravity to irrigate. This particular system is the most common and least restricted system. In most states, this system does not require construction permits. This system is often characterized as Laundry to Landscape (L2L). The system relies on valves, draining to a mulch basin, or the area of irrigation for certain landscape features (a mulch basin for a tree requires 12.6 ft^2). The drip system must be calibrated to avoid uneven distribution of grey water or overloading.

Delivery Methods

- Above Grade Gravity-Fed Irrigation.

Often used for washing machine or tub with gravity-fed irrigation systems:

- Grey water leaves source and passes through three-way diversion valve.
- Water enters either gully trap or surge tank.
- Tank water enters filter bag.
- Filtered water falls to irrigation system.
- Submersible Pump Diversion to Irrigation.

Often used for low level of contamination in water, providing that unsuitable water enters sewer:

- Grey Water discharges over a screen filter which can be diverted to a gully with a valve leading to the sewer.
- Screen filtered water enters a submersed pump.
- Pump sends water to irrigation system.
- Above Grade Gravity-Fed Pumped Irrigation.

Often used for laundry or bathroom grey water systems with pumped irrigation:

- Grey water leaves through three-way diversion valve.
- Enters surge tank.
- Tank water enters filter bag.
- Filtered water enters submersible pump.
- Pump supplies water to irrigation system.

Indoor Reuse

Recycled grey water from domestic appliances can be also be used to flush toilet. It application is based on standards set by plumbing codes. Indoor grey water reuse requires an efficient cleaning tank for insoluble waste, as well as a well regulated control mechanism.

The Uniform Plumbing Code, adopted in some U.S. jurisdictions, prohibits greywater use indoors.

Construction of Grey Water Systems

A written construction permit is required prior to building, fixing, relocating, or altering any system that *requires* a permit. Permits require usage data, member sizing, and soil conditions (to reduce pooling of grey water). Different states and regions are subject to different requirements of plans and system specifications. Standards set by state dictate codes for grey water systems based on definition of grey water, system type, and permit requirement. In many cases, construction can be done without contracting a professional.

Testing and Inspection

A grey water system that requires a permit must accurately determine the absorptive capacity of soils. Certain tests may be required for irregular soils, or undocumented areas.

A permitted system is required to be inspected. During construction the system is to be left uncovered. Additionally, implementation of certain devices can be prescribed in codes and requirements. A Reduced Pressure (RP) backflow device at property entry makes it easier to assess any amounts of crossover between grey water and a potentially potable water supply.

Challenges

The biggest concern in Integrated Urban Water Management systems, specifically grey water reuse is the introduction of untreated water into a landscape. It is important that grey water systems maximize natural purification through healthy topsoil, avoiding contact between greywater before and after filtration. Untreated grey water has the capacity to meet runoff and ultimately pollute water systems.

Research suggests that evaporated grey water can leave microorganisms that can be harmful if breathed or consumed. It is best practice to not use greywater in a sprinkler system, for this reason. Direct application of grey water can leave the aforementioned microorganisms on foliage. Grey water should not be used on fruits or vegetables (unless applied very carefully and specifically to the roots – although most state codes won't allow this).

Excess grey water which may not percolate into the soil could become runoff often leading the untreated water to waterways. Grey water should not be applied to saturated soil, and should be used conservatively.

Many household cleaners contain ingredients which can't be removed by a typical grey water filtration system. This requires an assessment of which substances will alter the quality of the grey water. Additionally, not all household appliances should be used for grey water recycling. Some should be diverted separately to sewage.

Grey water filtration systems are not equipped to handle the highest levels contaminants. Discretion is required as to when to use the grey water system. This includes consideration of what appliances to connect, as well as how much water is being processed at any given time.

One of the most significant challenges for IUWM could be securing a consensus on the definition of IUWM and the implementation of stated objectives at operational stages of projects. In the developing world there is still a significant fraction of the population that has no access to proper water supply and sanitation. At the same time, population growth, urbanization and industrialization continue to cause pollution and depletion of water sources. In the developed world, pollution of water sources is threatening the sustainability of urban water systems. Climate change is likely to affect all urban centers, either with increasingly heavy storms or with prolonged droughts, or perhaps both. To address the challenges facing IUWM it is crucial to develop good approaches, so that policy development and planning are directed towards addressing these global change pressures, and to achieving truly sustainable urban water systems.

STORM WATER MANAGEMENT

Stormwater is precipitation such as rain, sleet, or melting snow. In a natural setting only a small percentage becomes surface runoff, but as development occurs this percentage increases. This runoff usually flows into the nearest stream, creek, river, lake, or wetland.

Stormwater management involves the control of that surface runoff. The volume and rate of runoff both substantially increase as land development occurs. Construction of impervious surfaces, such as roofs, parking lots, and roadways, and the installation of storm sewer pipes which efficiently collect and discharge runoff, prevent the infiltration of rainfall into the soil. Management of stormwater runoff is necessary to compensate for possible impacts of impervious surfaces such as decreased groundwater recharge, increased frequency of flooding, stream channel instability, concentration of flow on adjacent properties, and damage to transportation and utility infrastructure. It is also now known that non-point source pollution washed off from those impervious surfaces is the 3rd leading cause of impairment to Pennsylvania's streams.

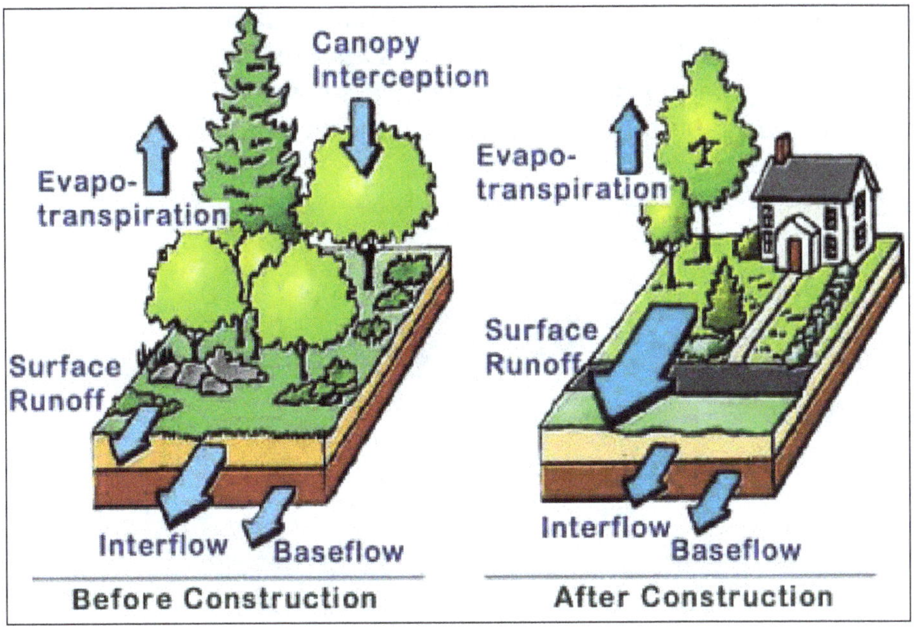

Hydrologic cycle showing stormwater runoff.

Why is Stormwater Management Important?

Stormwater management prevents physical damage to persons and property from flooding, and also prevents polluted run-off from negatively impacting local waterways. The installation of impervious surfaces interrupts the natural hydrologic cycle, and causes less infiltration, interception, and evapotranspiration than was present before any development occurred. Therefore, the volume and rate of flow of stormwater produced by the land surface have been greatly increased. The result of this larger amount of stormwater runoff significantly contributes to flooding, sediment deposition, erosion, nonpoint source pollution and stream channel instability.

Stormwater should be considered a resource that provides benefits such as groundwater recharge, which maintains flows in streams.

Stormwater management also reduces the frequency and severity of flooding. Traditional stormwater management takes surface runoff and diverts it to a detention pond, which holds the water and releases it at a constant rate over time. This approach allows the water to be returned to the watercourse at a high volume over a longer period of time, which does not necessarily rectify the problem and may actually create another. If stormwater is recharged into the groundwater, it can protect against erosion, flooding, and water quality degradation

Stormwater runoff is classified as one of the major sources of non-point source pollution impairment to waterways within Pennsylvania. According to the Pennsylvania Department of Environmental Protection (DEP), through September 2003 of the 82% of PA's streams assessed for water quality impairments, stormwater runoff has been documented as impairing over 1,700 miles of stream. The number of miles of impaired waters could rise once all the streams have been assessed or new reaches of streams become degraded by uncontrolled stormwater. Non-point source pollution is defined as pollution that originates from diffuse sources, and one cannot physically point to the exact, specific source.

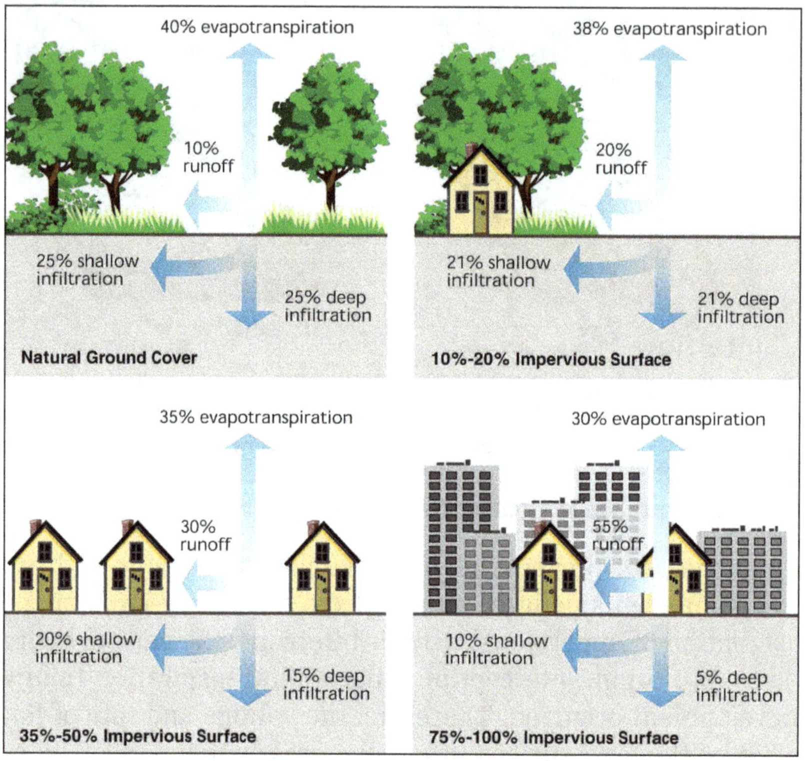

Stormwater runoff shown for percentages of impervious surfaces.

Common Non-point Source Pollution from Stormwater Runoff

Sediment

- Sediment constitutes the largest pollutant load associated with stormwater runoff in an urban setting. The loadings are exceptionally high in the case of construction activity without erosion controls.
- Sediment increases turbidity, harms aquatic and benthic habitat, and reduces capacity of impoundments.
- A number of other pollutants often attach to, and are carried by, sediment particles.

Nutrients

- The nutrients in stormwater runoff are phosphorus and nitrogen.

- In surface waters nutrient loads can lead to heavy algae growth, (especially in impoundments), and low dissolved oxygen levels.
- Nutrients enter the urban system in runoff from lawn and garden fertilizers (commercial and home), leaks from sanitary sewers and septic systems, and animal wastes.

Organic Matter

- Various forms of organic matter may be carried by stormwater in urban areas. Decomposition of this material by organisms in surface waters depletes oxygen levels.
- Low levels of dissolved oxygen severely impact water quality and life within surface waters.
- Sources of organic matter include leaking septic systems, garbage, and yard waste.

Bacteria

High bacterial levels in stormwater ru systems, garbage and animal waste.

IRRIGATION MANAGEMENT

Irrigation water management is the act of timing and regulating irrigation water applications in a way that will satisfy the water requirement of the crop without the waste of water, soil, plant nutrients, or energy. It means applying water according to crop needs in amounts that can be held in the soil available to crops and at rates consistent with the intake characteristics of the soil and the erosion hazard of the site.

Management is a prime factor in the success of an irrigation system. Large quantities of water, and often large labor inputs, are required for irrigation. The irrigator can realize profits from investments in irrigation equipment only if water is used efficiently. The net results of proper irrigation water management typically:

- Prevent excessive use of water for irrigation purposes.
- Prevent irrigation induced erosion.
- Reduce labor.
- Minimize pumping costs.
- Maintain or improve quality of ground water and downstream surface water.

- Increase crop biomass yield and product quality.

Tools, aids, practices, and programs to assist the irrigator in applying proper irrigation water management include:

- Applying the use of water budgets or balances to identify potential water application improvements.
- Applying the knowledge of soil characteristics for water release, allowable irrigation application rates, available water capacity, and water table depths.
- Applying the knowledge of crop characteristics for water use rates, growth characteristics, yield and quality, rooting depths, and allowable plant moisture stress levels.
- Water delivery schedule effects.
- Water flow measurement for on field water management.
- Irrigation scheduling techniques.
- Irrigation system evaluation techniques.

References

- Water-management-importance: omicsonline.org, Retrieved 14 June, 2020
- Barton, A.B. (2009). "Advancing IUWM through an understanding of the urban water balance". Australia's Commonwealth Scientific and Industrial Research Organisation (CSIRO). Retrieved 2009-09-14
- Water-resources-management, earth-and-planetary-sciences, topics: sciencedirect.com, Retrieved 15 July, 2020
- Behzadian, K; Kapelan, Z (2015). "Advantages of integrated and sustainability based assessment for metabolism based strategic planning of urban water systems". Science of the Total Environment. 527-528: 220–231. Bibcode:2015scten.527..220B. Doi:10.1016/j.scitotenv.2015.04.097. Hdl:10871/17351. PMID 25965035
- Stormwater-handbook, uploads: epcamr.org, Retrieved 16 August, 2020

5

Water Treatment

The process used to improve the quality of water to make it potable for various uses is referred to as water treatment. Water treatment is done through the processes of primary, secondary and advanced treatment. This chapter closely examines these processes used in water treatment for an extensive understanding of the subject.

WATER RECYCLING

All water is recycled water. Chances are, most, if not all of the molecules of water that you drink have passed through the bodies of other humans or animals. In addition, all water on the earth has been recycled countless times through the natural "hydrologic cycle." Precipitation in the form of snow or rain falls to the ground and then eventually flows to rivers, lakes and the ocean. The water then evaporates, rises into the sky, condenses into clouds, and falls back to the earth as rain or snow, completing the cycle. Some rain percolates into underground aquifers where it remains until feeding a surface water body at a lower elevation or being pumped out for use.

Although our sources of fresh water are finite, they are part of a naturally-replenishing system. The problem – we tend to use up the water faster than it can be replenished. The solution – use water more efficiently. Part of that is using less water through a variety of means, including more efficient irrigation systems, ocean friendly gardens that don't require much water and water-conserving appliances and shower heads. But an equally important part of the solution is reusing (and reusing, and reusing) the water we have.

Much of the water ("graywater") that goes down the drains from showers and sinks in our houses is suitable to be used locally for irrigation purposes. The rest ("black" water) can be used for irrigation after going through a conventional primary/secondary sewage treatment process, typically followed by filtration and disinfection. This treated wastewater is often referred to as reclaimed water and can be used to irrigate parks, golf courses, cemeteries and other landscaping. And 100% recycling is possible. Cape Coral, Florida recycles all of its wastewater during some years.

A few years ago the city launched a half-billion dollars worth of water and sewer treatment plant expansions to treat wastewater and pump it back through the city's irrigation system. This protects their waterways and assures the city has a reliable water source all year.

Other progressive communities that are encouraging capture, treatment and reuse of water onsite are San Francisco and the Yakima Basin Region of Washington.

Understanding and implementing water recycling requires a paradigm shift from thinking of "used" water as wastewater (something to get rid of as cheaply as possible) to thinking of it as a valuable resource. Existing, proven technology can be used to convert water that has previously been flushed down your toilet into drinking water that meets all applicable water quality standards and is probably purer than most expensive bottled water.

An example of just such a facility is the Groundwater Replenishment System in Orange County, California. There, about 100 million gallons per day of wastewater that has gone through primary and secondary treatment at Orange County Sanitation District's "Plant 1" is piped to a facility next door operated by Orange County Water District. The water is then processed through three additional purifying steps, using microfiltration, reverse osmosis, and advanced oxidation. At that point the water is pumped about 15 miles to spreading basins where the water is allowed to percolate through natural sand into underground aquifers. Wells then draw water out of the aquifer to be used as a fresh water supply. That system can be thought of as a continuous loop or a "man made" water cycle.

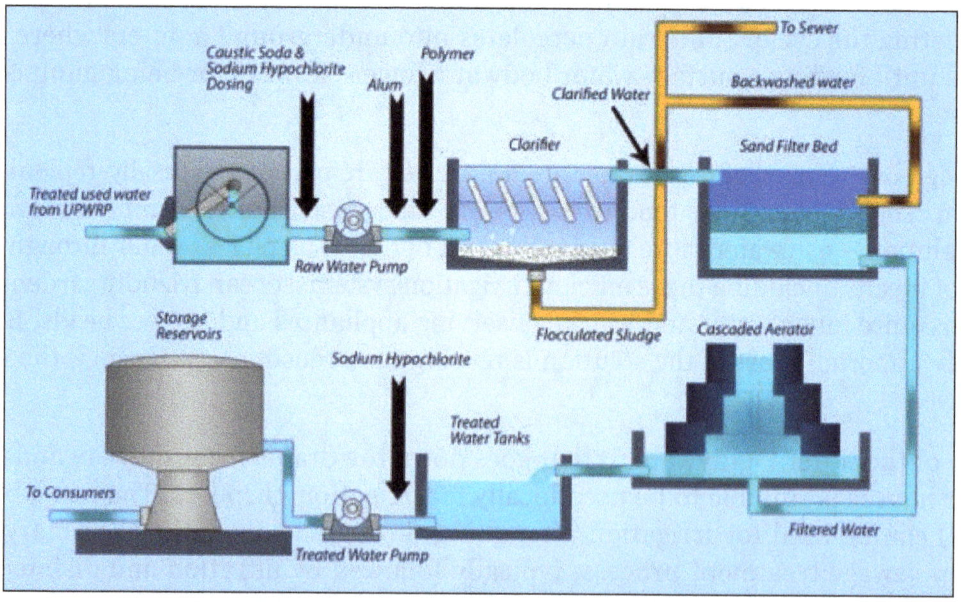

Following Orange County's lead, Los Angeles, San Diego and Ventura are in the process of expanding their reclaimed water systems and are evaluating the implementation of further treatment to allow potable use. In San Diego this is being termed "reservoir augmentation" or "indirect potable reuse" and has been branded Pure Water San Diego. And this isn't just a California thing. Since 1978, the Upper Occoquan Service Authority has been discharging recycled water into a stream above Occoquan Reservoir, a potable water supply source for Fairfax County, Virginia. Many cities pump their water supplies from rivers that receive treated wastewater from upstream cities. Although additional treatment processes such as filtration and chlorination are used to ensure that such water sources are safe, the beauty of the new "advanced" treatment processes such as the ones used by Orange County Water District is that they utilize several technologies to truly "re-purify" the water.

In Wichita Falls, Texas, drought conditions spurred officials to fast-track a temporary wastewater recycling project. It worked so well, producing just under 2 billion gallons of water in one year, that the city is now moving forward with construction of a permanent system.

In Florida, the Fiesta Village Wastewater Treatment Plant in Lee County processes wastewater for the South Ft. Myers and Cypress Lakes areas to be either used for reclaimed water or put back into nature to become part of the water cycle. This facility received the Florida Water Environment Association's 2011 Earle B. Phelps Award in the advanced wastewater treatment plant category. The award honors outstanding wastewater treatment plants in Florida that have maintained the highest levels of pollutants removal. The 5-mgd Fiesta Village AWTP, owned and operated by Lee County Utilities, has won the Phelps Award ten previous times.

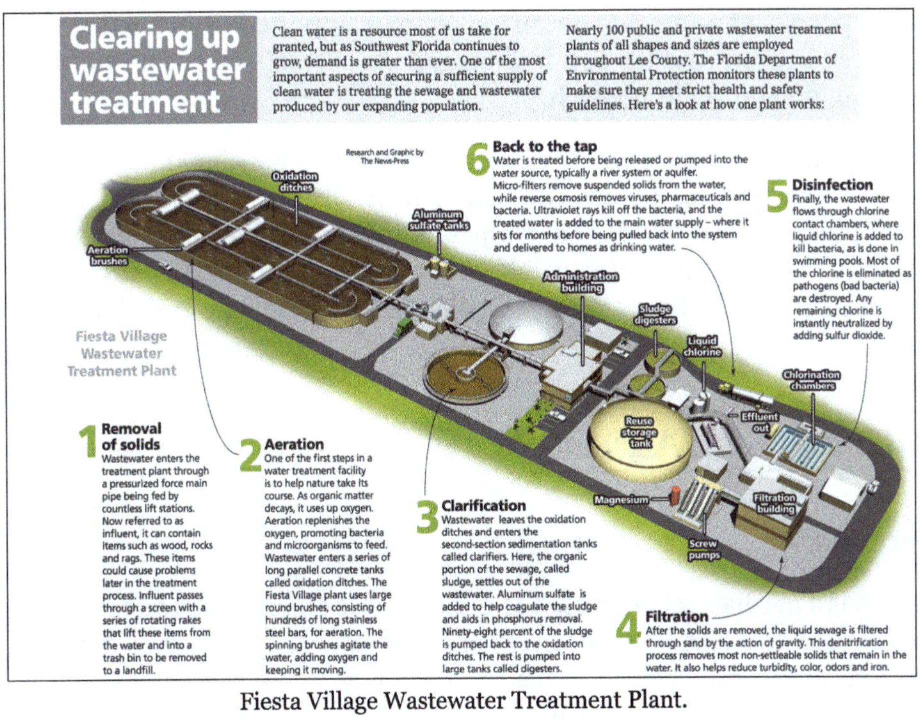

Fiesta Village Wastewater Treatment Plant.

One of the impediments to implementation of wastewater recycling systems, especially those that involve direct or indirect potable reuse of the treated wastewater, has been concern that these projects might represent a risk to public health. In fact, a 1998 National Research Council recommendation was that reclaimed water be used in drinking supplies only as "an option of last resort." Well, times have changed and the scientific understanding of the reliability of treatment processes has advanced. In early 2012 the National Research Council released a new report which says that reclaimed water can contribute a growing portion of the nation's drinking water supplies and be as safe as conventional sources. "We can really say that there is no difference from the risk standpoint," said Jorg Drewes, a water reuse expert who was on the panel. "You can have a supply that is as safe as the current drinking water supplies." As Ben Grumbles says in a NY Times article about the report "In essence, there is no wastewater, just wasted water." In 2015 WateReuse, American Water Works Association, Water Environment Federation and National Water Research Institute released a report Framework for Direct Potable Reuse which provides a context for Direct Potable Reuse (DPR), including the costs, benefits, energy requirements, and comparative issues with other water sources and measures. Following this introduction, three key components of a DPR program are examined: (1) regulatory considerations (e.g., measures to mitigate public health risks); (2) technical issues related to the production of advanced treated water; and (3) public support and outreach. This technology is truly coming of age.

Water recycling simultaneously addresses both a water supply problem and a waste disposal/ocean pollution problem. That 100 million gallons per day of wastewater in Orange County that is now being converted into high purity drinking water is 100 million gallons per day of water that doesn't need to be imported from northern California or the Colorado River. And it's 100 million gallons of wastewater that no longer flows into the ocean off Huntington Beach.

Water recycling is also feasible at the community or neighborhood level. In fact, small local systems may have advantages over large centralized systems by avoiding the use of long and costly conveyance systems (pipes and pumps). The main concept is that water is a natural resource that is too valuable to use just once and then throw away. Stop the Cycle of Insanity and Know Your H_2O. Nature has recycled water since time began.

PRIMARY TREATMENT

Bar Screens

A bar screen is a mechanical filter used to remove large objects, such as rags and plastics, from wastewater. It is part of the primary filtration flow and typically is the first, or preliminary, level of filtration, being installed at the influent to a wastewater treatment plant. They typically consist of a series of vertical steel bars spaced between 1 and 3 inches apart.

Bar screens come in many designs. Some employ automatic cleaning mechanisms using electric motors and chains, some must be cleaned manually by means of a heavy rake. Items removed from the influent are called screenings and are collected in dumpsters and disposed of in landfills. As a bar screen collects objects, the water level will rise, and so they must be cleared regularly to prevent overflow.

Grit Chamber

Grit is the heavy mineral material found in the wastewater influent in the treatment plant. It consists of gravel, sand, broken glass, fragments of metal and inorganic solids which have subsiding velocities or specific gravities greater than organic putrescible solids present in wastewater. Grit chambers are used to remove grit present in the wastewater. Grit has a specific gravity ranging from 2.4 to 2.65. They settle much more rapidly than organic solids.

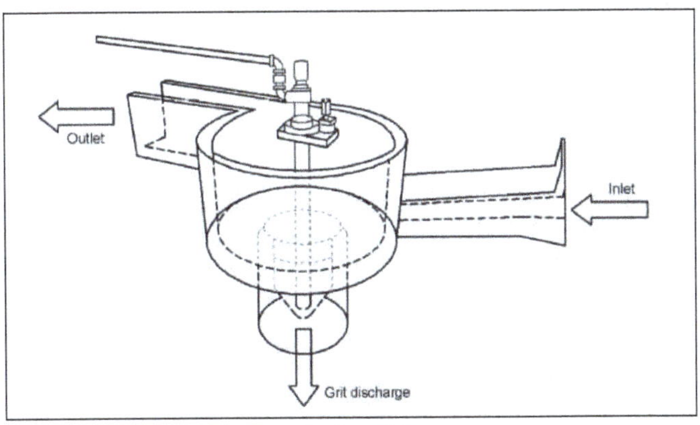

Grit chamber.

Grit chambers are used to remove grit present in the wastewater. Their functions are:

- To protect mechanical equipments used in the wastewater treatment plant from abrasion.
- To prevent heavy deposits in pipelines and channels.
- To reduce the frequency of digester cleaning.

- To reduce maintenance cost in high speed centrifuges which needs almost all grit particles to be removed.

It is also inevitable to reduce grit load for the safe operation of heat exchangers and pressure diaphragm pumps.

Classification of Grit Chambers

Grit chambers are classified into two, based on the mechanism of removal of grit. They are:

- Horizontal flow grit chambers.
- Aerated grit chambers.

Horizontal Flow Grit Chambers

It is a narrow open channel of about 10-18 meters long and with a depth of 1 to 1.3 m. The velocity in the grit chamber is maintained in such a way that the velocity will carry most organic particles through the chamber and will tend to re-suspend any that settle, but will permit the settling of heavier grit particles. So it is usually designed to maintain a velocity of 0.3 m/s. To maintain a fairly constant velocity of flow, a control section is used. These control sections are classified as following:

- Proportional flow weirs.
- Parshall flumes.
- Palmer-Bowlus flumes.
- Sutor weirs.

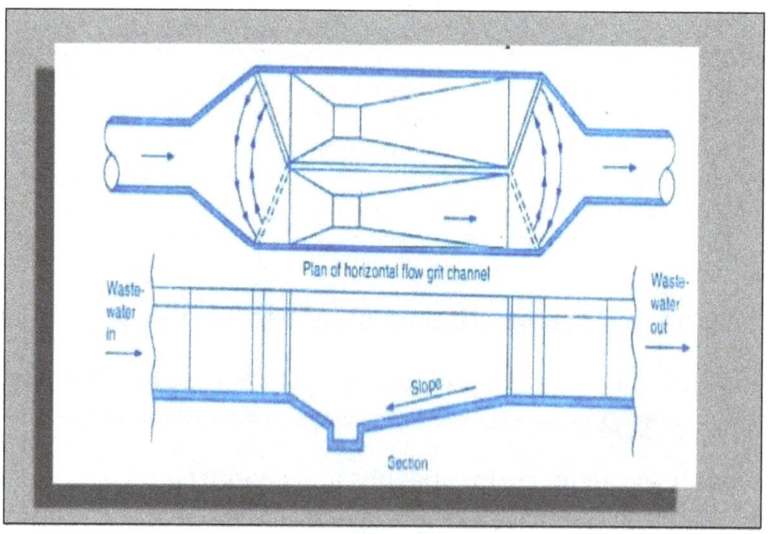

Horizontal flow grit chamber.

Aerated Grit Chambers

Aerated grit chambers consist of a spiral flow aeration tank enabled with air diffusion tubes placed on one side of the tank. They control the separation of inorganic and organic particles by producing a rolling flow pattern. The heavier particles tend to settle down to the flow due to their higher settling velocities and lighter particles are carried with the roll of spiral motion and eventually out of the tank.

Grit chambers are also classified on the basis of mechanism of cleaning them. They are:

- Mechanically cleaned grit chambers.
- Manually cleaned grit chambers.

If the influent flow exceeds 10 MLD, mechanized grit removal units are usually used.

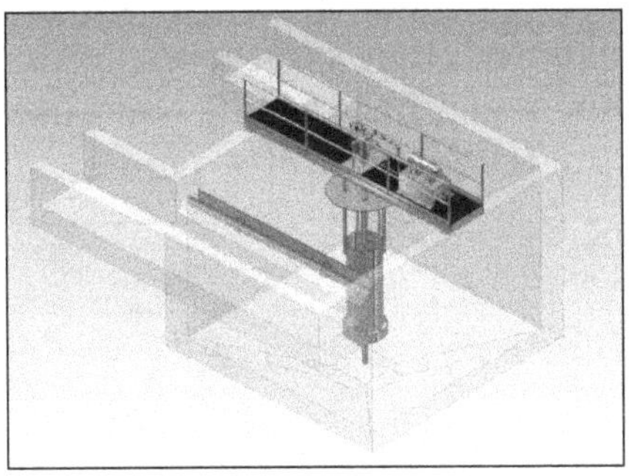

Aerated grit chamber.

Disposal of Grit

Grit is usually washed after removing it from the chamber, unless it may contain considerable amount of organic matter which causes unpleasant odor. It usually devoid of odor and resembles with sand and gravel particles after washing them. They are mainly disposed of by dumping or by a sanitary landfill. The method of disposal is selected on the basis of physical and chemical characteristics of grit, quality, availability of land for disposal etc.

Detritus Tank

A detritus tank is a grit removal unit which removes silt and some organic matter along with the grit. This is achieved by reducing the velocity of flow through the tank and increasing the detention time. Thus it is mainly used for removing finer particles than those removed by a grit chamber. It is a continuous flow settling tank which is either of in rectangular or in a square shape.

The sides of tanks are vertical which are gradually tapered to the bottom to form a trough for the collection of detritus. The velocity of flow maintained in the tank will be of 0.2 to 0.3 m/s and the detention time will be 3-4 minutes. The depth of tank will vary from 2.5 m to 3.5 m.

Detritus tank.

Skimming Tank

Skimming tanks are used to remove floating substances like grease, oil, fats, soap, fruit skins etc. if these are not removed, they can cause severe damages to several equipments in a wastewater treatment plant. Skimming tanks used are generally rectangular or circular in shape having a detention time of 3-5 minutes.

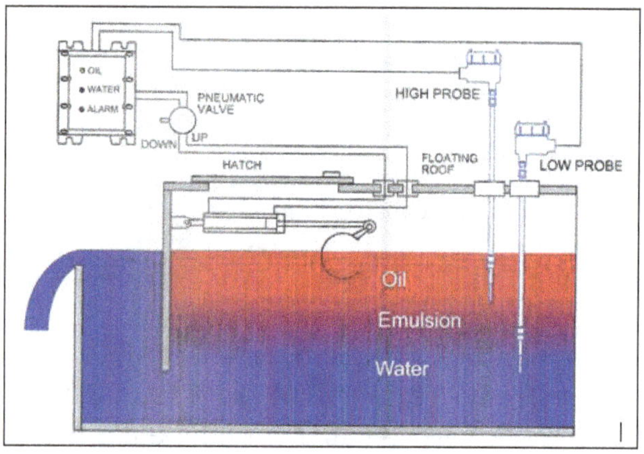

Oil Skimming Tank.

When the flow enters the skimming tank, these floating matters will rises and remains in the surface of wastewater and subsequently removed. The clear wastewater is removed either through an outlet places in the bottom of tank or through partitions given at the bottom sidewalls.

SECONDARY TREATMENT

Aeration Basin

An aeration basin (also called an aeration lagoon) is a holding and/or treatment pond provided with artificial aeration to promote the biochemical oxidation of wastewaters.

There are many other biological processes for treatment of wastewaters, for example the activated sludge process, trickling filters (also called biofilters) and rotating biological contactors. They all have in common the use of oxygen (or air) and microbial action to biotreat the pollutants in wastewaters.

Aeration basin using a submerged grid of air diffusers.

Types of Aerated Basins

There are many methods for aerating a lagoon or basin:

- Motor-driven floating surface aerators.
- Motor-driven submerged aerators.
- Motor-driven fixed-in-place surface aerators.
- Injection of compressed air through submerged diffusers.
- Floating surface aerators.

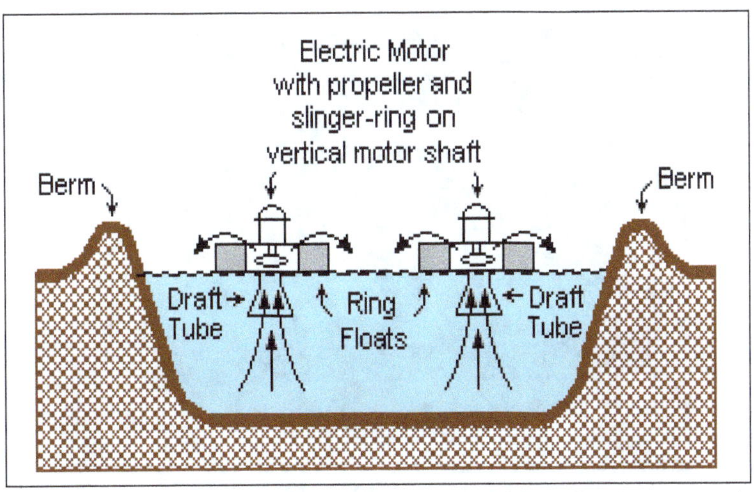

A Typical Surface-Aerated Basing (using motor-driven floating aerators).

Ponds or basins using floating surface aerators achieve 80 to 90% removal of BOD with retention times of 1 to 10 days. The ponds or basins may range in depth from 1.5 to 5.0 metres.

In a surface-aerated system, the aerators provide two functions: they transfer air into the basins required by the biological oxidation reactions, and they provide the mixing required for dispersing the air and for contacting the reactants (that is, oxygen, wastewater and microbes). Typically, the floating surface aerators are rated to deliver the amount of air equivalent to 1.8 to 2.7 kg O_2/kWh. However, they do not provide as good mixing as is normally achieved in activated sludge processes and therefore aerated basins do not achieve the same performance level as activated sludge units.

Biochemical oxidation processes are sensitive to temperature and, between 0 °C and 40 °C, the rate of biological reactions increases with temperature. Most surface aerated vessels operate at a temperature of between 4 °C and 32 °C.

Clarifier

Three wastewater/sewage clarifiers at the 'Aikahi wastewater treatment plant in Hawaii. They appear to have a floating cover to reduce the odor because the plant is very close to a residential area.

Circular clarifier with surface skimmer

Clarifiers are settling tanks built with mechanical means for continuous removal of solids being deposited by sedimentation. A clarifier is generally used to remove solid particulates or suspended solids from liquid for clarification and (or) thickening. Concentrated impurities, discharged from the bottom of the tank are known as sludge, while the particles that float to the surface of the liquid are called scum.

Applications

Pretreatment

Before the water enters the clarifier, coagulation and flocculation reagents, such as polyelectrolytes and ferric sulfate, can be added. These reagents cause finely suspended particles to clump together and form larger and denser particles, called flocs, that settle more quickly and stably. This allows the separation of the solids in the clarifier to occur more efficiently and easily; aiding in the conservation of energy. Isolating the particle components first using these processes may reduce the volume of downstream water treatment processes like filtration.

Potable Water Treatment

Water being purified for human consumption, is treated with floculation reagents, then sent to the clarifier where removal of the flocculated coagulate occurs producing clarified water. The clarifier works by permitting the heavier and larger particles to settle to the bottom of the clarifier. The particles then form a bottom layer of sludge requiring regular removal and disposal. Clarified water then proceeds through several more steps before being sent for storage and use.

Waste Water Treatment

Sedimentation tanks have been used to treat wastewater for millennia.

Primary treatment of sewage is removal of floating and settleable solids through sedimentation. Primary clarifiers reduce the content of suspended solids and pollutants embedded in those suspended solids. Because of the large amount of reagent necessary to treat domestic wastewater, preliminary chemical coagulation and flocculation are generally not used, remaining suspended solids being reduced by following stages of the system. However, coagulation and flocculation can be used for building a compact treatment plant (also called a "package treatment plant"), or for further polishing of the treated water.

Sedimentation tanks called secondary clarifiers remove flocs of biological growth created in some methods of secondary treatment including activated sludge, trickling filters and rotating biological contactors.

Mining

Methods used to treat suspended solids in mining wastewater include sedimentation and floc blanket clarification and filtration. Sedimentation is used by Rio Tinto Minerals to refine raw ore into refined borates. After dissolving the ore, the saturated borate solution is pumped into a large settling tank. Borates float on top of the liquor while rock and clay settles to the bottom.

Technology

Rectangular sedimentation tanks with effluent weir structure visible above the fluid surface.

Drained circular sedimentation tank showing central inlet baffles on the right with solids scraper and skimmer arms visible under the rotating bridge.

Although sedimentation might occur in tanks of other shapes, removal of accumulated solids is easiest with conveyor belts in rectangular tanks or with scrapers rotating around the central axis of circular tanks. Mechanical solids removal devices move as slowly as practical to minimize resuspension of settled solids. Tanks are sized to give water an optimal residence time within the tank. Economy favors using small tanks; but if flow rate through the tank is too high, most particles will not have sufficient time to settle, and will be carried with the treated water. Considerable attention is focused on reducing water inlet and outlet velocities to minimize turbulence and promote effective settling throughout available tank volume. Baffles are used to prevent fluid velocities at the tank entrance from extending into the tank; and overflow weirs are used to uniformly distribute flow from liquid leaving the tank over a wide area of the surface to minimize resuspension of settling particles.

Tube Settlers

Tube settlers are commonly used in rectangular clarifiers to increase the settling capacity by reducing the vertical distance a suspended particle must travel. High efficiency tube settlers use a stack of parallel tubes, rectangles or flat pieces separated by a few inches (several centimeters) and sloping upwards in the direction of flow. This structure creates a large number of narrow parallel flow pathways encouraging uniform laminar flow as modeled by Stokes' law. These structures work in two ways:

- They provide a very large surface area onto which particles may fall and become stabilized.
- Because flow is temporarily accelerated between the plates and then immediately slows down, this helps to aggregate very fine particles that can settle as the flow exits the plates.

Structures inclined between 45° and 60° may allow gravity drainage of accumulated solids, but shallower angles of inclination typically require periodic draining and cleaning. Tube settlers may allow the use of a smaller clarifier and may enable finer particles to be separated with residence times less than 10 minutes. Typically such structures are used for difficult-to-treat waters, especially those containing colloidal materials.

Tube settlers capture the fine particles allowing the larger particles to travel to the bottom of the clarifier in a more uniform form. The fine particles then build up into a larger mass which then slides down the tube channels. The reduction in solids present in the outflow allows a reduction in the clarifier footprint when designing. Tubes made of PVC plastic are a minor cost in clarifier design improvements and may lead to an increase of operating rate of 2 to 4 times.

Operation

In order to maintain and promote the proper processing of a clarifier, it is important to remove any corrosive, reactive and polymerisable component first, or any material that may foul the outlet stream of water to avoid any unwanted side reactions, changes in the product or cause damage to any of the water treatment equipment. This is done by routine inspections and the frequent cleaning of the quiescent zones and the inlet and outlet areas of the clarifier in order to ascertain the extent of sediment build up and to clean and remove any scouring, litter, weeds or debris that may have accumulated over time.

Water being introduced into the clarifier should be controlled to reduce the velocity of the inlet flow. Reducing the velocity maximizes the hydraulic retention time inside the clarifier for sedimentation and helps to avoid excessive turbulence and mixing; thereby promoting the effective settling of the suspended particles. To further discourage the overt mixing within the clarifier and increase the retention time allowed for the particles to settle, the inlet flow should also be distributed evenly across the entire cross section of the settling zone inside the clarifier, where the volume is maintained at 37.7 percent capacity.

The sludge formed from the settled particles at the bottom of each clarifier, if left for an extended period of time, may become gluey and viscous, causing difficulties in its removal. This formation of sludge, promotes anaerobic conditions and a healthy environment for the growth of bacteria. This can cause the resuspension of particles by gases and the release of dissolved nutrients throughout the water fluid, reducing the effectiveness of the clarifier. Major health issues and problems can also occur further down the track of the water purification system or hinder the health of the fish found downstream of the clarifier.

New Development

Improvements and modifications have been made to enhance clarifier performance dependent on the constraints of the substance undergoing the separation.

Addition of flocculants is common to aid separation in clarifiers, but density difference of flocculant concentrate may cause treated water to have an excessive flocculant concentration. Uniform flocculent concentration can be improved and flocculant dosage reduced by installation of an intermediate diffused wall perpendicular to the flow in the clarifier.

The two dominant forces acting upon the solid particles in clarifiers are gravity and particle interactions. Disproportional flow can lead to turbulent and hydraulic instability and potential flow short-circuiting. Installation of perforated baffle walls in modern clarifiers promotes uniform flow across the basin. Rectangular clarifiers are commonly used for high efficiency and low running cost. Improvements of these clarifiers were made to stabilize flow by elongation and narrowing of the tank.

References

- Wastewater-Recycling, beachapedia.org: Retrieved 17 January, 2020
- Streicher, Dennis. "Waste Water Treatment Plant Elmhurst, Illinois A Virtual Tour". Elmhurst College. Archived from the original on 29 March 2012. Retrieved 18 February 2012
- What-are-grit-chambers-classification-disposal-grit-skimming-tanks: civildigital.com, Retrieved 18 February, 2020
- Crittenden, John C.; Trussell, R. Rhodes; Hand, David W.; Howe, Kerry J.; Tchobanoglous, George (2012). MWH's water treatment: principles and design (3rd ed.). Hoboken, N.J.: John Wiley & Sons. ISBN 9780470405390
- Smart-easy-ways-to-reuse-wastewater: conserve-energy-future.com, Retrieved 19 March, 2020
- Anlauf, Harald (2003). "Mechanische Fest/Flüssig-Trennungim Wandel der Zeit". Chemie Ingenieur Technik. 75 (10): 1460–1463. Doi:10.1002/cite.200303283

Permissions

All chapters in this book are published with permission under the Creative Commons Attribution Share Alike License or equivalent. Every chapter published in this book has been scrutinized by our experts. Their significance has been extensively debated. The topics covered herein carry significant information for a comprehensive understanding. They may even be implemented as practical applications or may be referred to as a beginning point for further studies.

We would like to thank the editorial team for lending their expertise to make the book truly unique. They have played a crucial role in the development of this book. Without their invaluable contributions this book wouldn't have been possible. They have made vital efforts to compile up to date information on the varied aspects of this subject to make this book a valuable addition to the collection of many professionals and students.

This book was conceptualized with the vision of imparting up-to-date and integrated information in this field. To ensure the same, a matchless editorial board was set up. Every individual on the board went through rigorous rounds of assessment to prove their worth. After which they invested a large part of their time researching and compiling the most relevant data for our readers.

The editorial board has been involved in producing this book since its inception. They have spent rigorous hours researching and exploring the diverse topics which have resulted in the successful publishing of this book. They have passed on their knowledge of decades through this book. To expedite this challenging task, the publisher supported the team at every step. A small team of assistant editors was also appointed to further simplify the editing procedure and attain best results for the readers.

Apart from the editorial board, the designing team has also invested a significant amount of their time in understanding the subject and creating the most relevant covers. They scrutinized every image to scout for the most suitable representation of the subject and create an appropriate cover for the book.

The publishing team has been an ardent support to the editorial, designing and production team. Their endless efforts to recruit the best for this project, has resulted in the accomplishment of this book. They are a veteran in the field of academics and their pool of knowledge is as vast as their experience in printing. Their expertise and guidance has proved useful at every step. Their uncompromising quality standards have made this book an exceptional effort. Their encouragement from time to time has been an inspiration for everyone.

The publisher and the editorial board hope that this book will prove to be a valuable piece of knowledge for students, practitioners and scholars across the globe.

Index

www.ingramcontent.com/pod-product-compliance
Lightning Source LLC
LaVergne TN
LVHW080805220926
848485LV00011B/210